Hamlyn all-colour paperbacks

Iain Nicolson BSc

# Exploring
# the Planets

illustrated by James Nicholls

**Hamlyn · London**
Sun Books · Melbourne

## FOREWORD

This guide tells the story of the exploration of the planets and their satellites from the time when men first noticed their presence in the sky, to the present-day space probes and manned flights. The earliest watchers of the planets had to make all their observations with their eyes alone, and so the amount which has been discovered about the planets in the three hundred and fifty or so years which have passed since the invention of the telescope immeasurably exceeds all that was discovered up till that time since the dawn of history.

Nowadays we are entering a new phase of planetary exploration, just as revolutionary as the invention of the telescope – direct exploration. It is still true, though, that despite space probes the major part of space exploration is carried out by means of Earth-based instruments, optical and radio-telescopes and other, stranger equipment described in the text.

For a long time, little work was done on the planets by professional astronomers, who were more concerned with the universe as a whole. However, intensive research is now going on and developments are rapid, so rapid in fact that some information goes out of date very quickly. There is still scope for the amateur observer to make useful contributions to the exploration of the planets, and some of the ways in which this can be done are described.

I.K.M.N.

Published by The Hamlyn Publishing Group Limited
London · New York · Sydney · Toronto
Hamlyn House, Feltham, Middlesex, England
In association with Sun books Pty Ltd, Melbourne

Copyright © The Hamlyn Publishing Group Limited 1970

ISBN 0 600 00120 2
Phototypeset by Filmtype Services Ltd, Scarborough
Colour separations by Schwitter Limited, Zurich
Printed in Holland by Smeets Weert

# CONTENTS

The planets Jupiter (bright, off-white in colour) and Mars (reddish in colour) in the night sky

## INTRODUCTION

A casual glance at the night sky reveals a variety of interesting objects. The Moon is the dominant body, passing through its cycle of phases from New Moon to Full Moon and back again roughly once a month. The night sky is dotted with thousands of stars of varying brightness and colour, many of them grouped into easily recognized patterns or *constellations*. We now know that the stars are immense spheres of glowing gas, like our own Sun, but they lie at such enormous distances that they appear merely to be tiny points of light. Most nights it is possible to see one or two meteors (page 71) flash across the sky, and occasionally the ghostly form of a comet may be observed.

But there is another class of object to be seen. Whereas the stars seem to maintain fixed positions in the sky, there are a few star-like objects which seem to move slowly day by day,

4

month by month, against the background constellations. These are the planets, or 'wandering stars' as they were known to the ancients. The existence of planets has been known for many thousands of years, and five of these planets were known to the Egyptian and Babylonian observers of some five thousand years ago.

Nowadays we know that the planets are solid bodies like the Earth with no light of their own, shining by reflected sunlight. But the ancient astronomers, with no equipment but their own eyes, saw the planets simply as wandering stars and gave them the names of their mythological gods and goddesses. To give them their Latin names, the five planets were as follows: Mercury, the winged messenger of the gods; Venus, the goddess of love; Mars, the god of war; Jupiter, the king of the gods; and Saturn, father of Jupiter and known as 'Old Man Time'. The ancient astronomers observed the positions of these 'wandering stars' but found it hard to account for their motions.

Earth, Moon and Sun compared in size (distance not to scale)

## Ptolemy's universe

From around 700 BC to AD 200 the Greek civilization was at its height. The Greeks were perhaps the earliest true scientists and made many important advances in astronomical knowledge. Eratosthenes, for example, succeeded in calculating the circumference of the Earth quite accurately. However, the Greek view of the universe was summarized by Ptolemy in the second century AD. Ptolemy placed the Earth at the centre of the universe and the other bodies revolved around it in circular paths in the following order: the Moon, Mercury, Venus, the Sun, Mars, Jupiter and Saturn. Outside this lay the sphere of fixed stars. After Ptolemy's time the Greek civilization declined and no new ideas on the layout of the universe were to emerge for well over a thousand years.

## New ideas

It was apparent even in Ptolemy's time that the observed motions of the planets could not be wholly explained by assuming that they moved round the Earth in circular orbits, and he tried to meet this objection by superimposing additional motions on top of the basic circles.

In 1543 Copernicus (Nicolaus Koppernigk) published a revolutionary new theory in which he overcame many of the objections to the Ptolemaic system. He suggested that the Sun lay at the centre of the universe and that all the planets, Earth included, revolved about it. The Church at the time held the view that the Earth was the most important body in the universe and must therefore be in the centre, and so Copernicus' views were considered heretical. Nonetheless his theory began to catch on with thinking men.

The discrepancies still remaining between Copernicus' theoretical and the actual motions of the planets were tidied up at the beginning of the seventeenth century by Johannes Kepler. He suggested that the planets moved round the Sun not in circles but in elliptical orbits.

Around this time the telescope was invented by Lippershey in Holland, and in 1609 the Italian Galileo Galilei made the first systematic telescopic astronomical observations. He became convinced of the Copernican theory and was tried for his heretical views.

The Ptolemaic universe

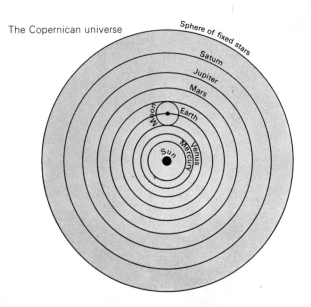

The Copernican universe

## Newton and gravitation

Why do the planets continue in their orbits around the Sun? The answer was supplied by one of the greatest men of science of all time, Isaac Newton. Born in 1642, he soon made a name for himself in nearly all the fields of science known in his time, particularly mathematics, physics and astronomy. He invented a new form of telescope, the reflector, which used mirrors instead of lenses to collect light.

It seems to be true that Newton was prompted to consider why bodies move as they do by watching an apple fall from a tree and wondering why it should fall at all. His theory of *universal gravitation* explained the planetary motion – and the falling of apples – by assuming that every body in the universe attracted every other with a force depending on the masses of the bodies and the inverse square of their separations. In other words, if the distance between two bodies were doubled, the attractive force would be quartered.

A Newtonian reflector made by Sir William Herschel

(*Above*) The discovery of Uranus by Herschel took the astronomical world by surprise.

(*Below*) The telescopic recognition of a planet. A planet can be detected either by its disc-like appearance or by its motion against the background stars.

Thus the planets and the Sun attract one another; the planets maintain their orbits around the Sun because their speed is just sufficient to prevent their falling into the Sun and not great enough to enable them to fly away.

Astronomy developed rapidly after Newton and larger and better telescopes were constructed. Sir William Herschel, with a telescope of his own construction, discovered in 1781 a completely new planet, Uranus, orbiting further out than Jupiter. This discovery took the astronomical world completely by surprise since for thousands of years it had been assumed that there were only five planets, apart from the Earth.

Calculations derived from the movements of Uranus (page 116) led to the discovery of yet another planet, Neptune, in 1846. While Uranus can just be seen with the naked eye, Neptune needs telescopic aid.

The outermost planet so far known, Pluto, was discovered in 1930 after a prolonged search based on calculations by Percival Lowell. Thus the solar system contains nine major planets. In addition, between Mars and Jupiter lie the minor planets.

The Sun. The dark areas are sunspots, the light areas faculae.

## A brief picture of our solar system

Let us look at our solar system as present-day astronomers see it. The Earth, as we know, is one of nine planets orbiting the Sun, which is itself nothing more than a very ordinary star. The Sun, however, is of vital importance to us, providing light and heat in just the right quantities to maintain life as we know it. Only a small change in the total output of energy from the Sun would be sufficient to wipe out our form of life.

The Sun is a ball of glowing gas, principally hydrogen, some 864,000 miles in diameter. It is well over a million times larger in volume than the Earth, but the mass of the Sun is only 330,000 times greater than that of the Earth. So the Sun's density is much lower, in fact 1·4 times that of water as against 5·5 for the Earth. The temperature of the Sun's surface (or photosphere) is about 6,000° Kelvin (the Kelvin, or Absolute, temperature scale starts at −273° Centigrade), whereas in the highly compressed central regions the temperature is thought to approach 20,000,000°. The Sun produces this continual and steady output of energy by operating just like

The Moon

a controlled hydrogen bomb. In the very hot central region, atoms of hydrogen are fused together to form the element helium, and large quantities of energy are given off in the process. Besides visible light and heat, the Sun gives out other sorts of radiation such as X-rays and radio waves.

The Sun's family consists principally of the nine major planets and their moons, and of course the thousands of minor planets. The Sun also carries with it a large amount of 'space debris' in the form of dust particles and tenuous gas, as well as the swarms of tiny meteors and the loose agglomerations of matter which make up comets.

The planets compared in size. (*Left to right*) Mercury, Venus, Earth, Mars, Jupiter, Saturn, Uranus, Neptune, Pluto

## Beyond the fringe

Beyond the fringes of the solar system lies nothing but an exceedingly thin mixture of gas and dust until we reach some other stars. The distances separating the stars are so great that the unit employed for measurement is the light-year—the distance travelled by light in one year. As the speed of light in space is 186,000 miles per second, a light year is equal to about six million million miles; yet even on this reckoning the nearest star, Proxima Centauri, is over four light-years away.

The stars vary considerably in size and temperature. Colour is quite a good guide to a star's temperature; red stars are relatively cool, while blue stars are exceedingly hot. The Sun is a yellow star. Stars are known which are smaller than the Earth, while others are so enormous that if they were placed where the Sun is now, the Earth would be well inside them.

Stars are often found grouped into clusters which may number from a few dozen to tens of thousands of stars in the case of the compact globular clusters. The Sun, stars and star clusters alike are all part of a giant system of something like a hundred thousand million stars, known as the Galaxy. This is disc-shaped, with stars more closely congregated towards the centre and spiral arms composed of stars, gas and dust, radiating from it. The Sun lies on one of the spiral arms.

The Galaxy itself is not alone however. At least a thousand

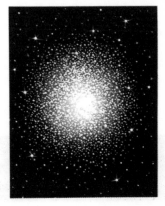

(*Above*) The position of M31, the Great Nebula in Andromeda, near the square of Pegasus
(*Right*) A globular cluster

A spiral galaxy

million other galaxies can be seen in the largest telescopes. Galaxies, too, tend to form into groups and clusters. The local group, of which our Galaxy is a member, numbers twenty-seven, one of which is the Great Andromeda Galaxy lying at a distance of about two million light-years, whose position in the sky is shown on the opposite page.

The galaxies seem to be moving apart at great speeds and this observation leads to two theories of the origin of the universe, the 'big bang' theory whereby all the matter in the universe was formed at one time and is steadily moving apart, and the 'steady state' theory whereby matter is continually formed between the moving galaxies. The origin of the universe remains a mystery.

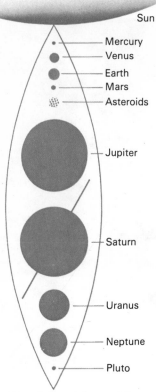

- Sun
- Mercury
- Venus
- Earth
- Mars
- Asteroids
- Jupiter
- Saturn
- Uranus
- Neptune
- Pluto

(*Above*) Jeans' theory of the origin of the solar system
(*Below*) Woolfson's theory

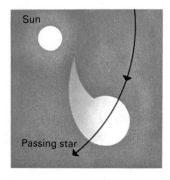

Sun

Passing star

## Origin of the solar system

Though the origin of the universe as a whole remains unsolved, we might expect the problem of the formation of the solar system to be readily answered, but the situation here is just as confused and no one theory is entirely satisfactory.

Ancient mythologies had many accounts of the origin of the Earth and planets, but the first truly scientific theory was suggested in 1796 by the French mathematician Laplace. He supposed that the Sun and planets began as a slowly rotating gas cloud which began to shrink and to rotate faster and faster. Eventually rings of matter were thrown off the edge of the cloud and the central part contracted to form the Sun. The rings of matter were supposed to form planets.

One of the objections to this theory was the question of angular momentum. This is a quantity which, roughly speaking, measures the amount of rotational energy that a body possesses, and it is a fundamental law that angular momentum can be transferred but not destroyed. At the present time, nearly all the angular momentum of the

The nebular hypothesis

solar system is concentrated in the planets, whereas for a contracting cloud like Laplace's most of the angular momentum should be possessed by the Sun.

Sir James Jeans overcame the angular momentum problem by suggesting that another star passing close to the Sun drew a cigar-shaped filament of matter from it and set it in rotation around the Sun. From this filament, planets were supposed to form, with the largest planets in the middle of the 'cigar' and the smallest (Mercury and Pluto) at the tapered ends. There were serious drawbacks to this hypothesis, though, and it too was eventually ruled out.

Most present-day theories are variants of the original Laplace hypothesis. The angular momentum problem is overcome by supposing that this quantity was transferred from the Sun to the cloud by means of some agency such as magnetic fields; that is, the Sun slowed down and the cloud speeded up. A tidal theory whereby the planetary material was drawn by the Sun from a passing star has been suggested by Professor M. M. Woolfson of York University and seems quite promising, but has not, as yet, gained much support.

# THE MOVEMENTS OF THE PLANETS

The nine planets and their attendant satellites revolve around the Sun at distances ranging from 36 million miles in the case of Mercury to about 3,666 million miles for Pluto, their orbital periods ranging from 88 days to 249 years respectively. Thus, although an average human being could attain an age of over three hundred 'Mercury years', even Methusela could not have lived to his first 'birthday' on Pluto!

All the planets move in the same direction and very nearly in the same plane. Thus the planets can always be found within a narrow band of sky, known as the *zodiac*, through the centre of which lies the apparent path of the Sun against the background stars, the *ecliptic*. The constellations lying in this region are known as the zodiac constellations and the positions of the planets in these constellations used to be of great importance to astrologers for the purpose of 'predicting' future events. It must be emphasized that astronomy is a pure science and in no way connected with astrology (though in ancient and mediaeval times the two were closely intertwined).

The planets appear to advance slowly through the zodiac

The planets of the solar system. Pluto is not shown here as its orbit lies for part of the time outside the orbit of Neptune, and part of the time inside.

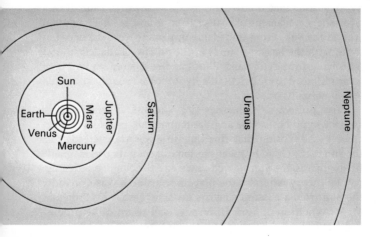

constellations, the rate of advance depending upon the orbital periods of the planets concerned. However the apparent motions are not uniform. The normal direction of motion of a planet is called *direct motion*. The planets moving outside the Earth's orbit – and thus more slowly – will at certain times appear to reverse their motion and carry out *retrograde motion*. Why this happens is explained for the case of Mars in the diagram. The apparent motion of Mars in the sky is shown together with the relative positions of the Earth and Mars. From positions 1 to 3, Mars moves in the ordinary direct fashion, while between 3 and 5, the Earth is actually overtaking Mars and Mars appears to move backwards. Thereafter Mars seems to resume its normal motion. Positions 3 and 5 where the planet reverses direction are called stationary points. A good example of this sort of behaviour is given by watching the motion against a distant background of a car which you happen to be overtaking.

This diagram shows why the outer planets appear to reverse their direction in the sky.

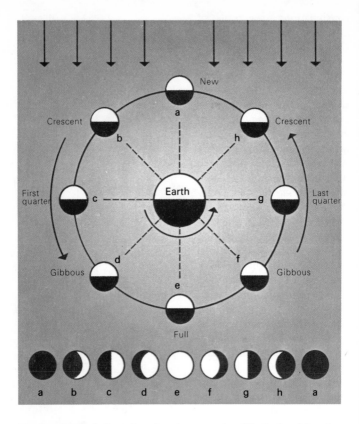

Kepler showed that the planets move in elliptical orbits. An ellipse is rather like a squashed circle and is said to have two foci; a circle has only one focus, the centre. An easy way of drawing an ellipse is to stick two pins into a piece of paper, one at each focus, pass a loop of string over the pins, draw the string tight with a pencil and, keeping it tight, draw round the pins. (It is a law of nature that one of the pins will always fall down.) The further apart the foci, the longer and narrower will be the ellipse, and a measure of the shape of an ellipse is given by the eccentricity. The higher the eccentricity, the narrower the ellipse. A circle is simply an ellipse with eccentricity zero. The eccentricity of planetary orbits varies from

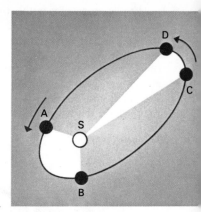

(*Opposite*) This illustrates how the phases of the Moon occur. The corresponding lower illustrations show how the Moon appears from Earth.

(*Right*) Kepler's second law. The line between the Sun and a planet sweeps out equal areas of space in equal times. Thus the area SAB equals the area SCD.

0·007 (Venus), to 0·248 (Pluto), while comets tend to have highly eccentric orbits.

Kepler put forward three laws of planetary motion as follows:

1. Planets move around the Sun in elliptical orbits with the Sun at one focus of the ellipse.
2. The line between a planet and the Sun (the radius vector) sweeps out equal areas of space in equal time. Thus a planet moves fastest when nearest the Sun (*perihelion*) and slowest when farthest away (*aphelion*).
3. The square of the orbital period depends on the cube of a planet's distance from the Sun. Thus, if the distance of one planet is known, all the others can be calculated if we know their orbital periods.

As the Moon travels round the Earth, it goes through a complete cycle of phases from *new*, when it is completely dark, and close to the Sun in the sky, to *full*, when its visible hemisphere is completely illuminated and it is opposite the Sun in the sky, and back again. Why this is so is readily seen from the diagram.

The planets Mercury and Venus move round the Sun in orbits within the Earth's and show phases like the Moon. When either of these planets is on the far side of the Sun, its illuminated hemisphere is turned towards us, whilst when it is roughly between us and the Sun, its dark side faces the Earth.

The planets moving within the Earth's orbit, that is, Mercury and Venus, are called the *inferior planets,* whilst those outside the Earth's path are the *superior planets.* When a planet lies in line with the Sun as seen from the Earth, it is said to be *in conjunction.* Now, the inferior planets come into conjunction at two points in their orbits—once when they come between the Earth and the Sun, *inferior conjunction,* and again when they are on the far side of the Sun, *superior conjunction.* The

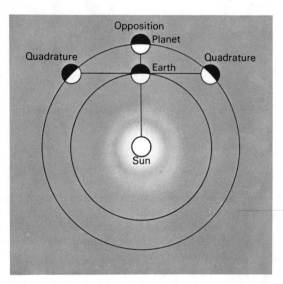

Opposition and quadrature of a planet

superior planets can only come into superior conjunction since they cannot pass between the Earth and the Sun.

When one of the inner planets is at its greatest apparent distance from the Sun, the angle between the Earth, the planet and the Sun is ninety degrees. This position is called *greatest elongation,* east or west depending on whether the planet appears to lie east or west of the Sun in the sky. At an eastern elongation the planet is visible in the evening sky, and at a western elongation it can be seen in the morning before sunrise. At greatest elongation a planet shows half phase – at this point the planet is said to be at *dichotomy*.

An outer planet lying on a line extended through the Sun and the Earth is *in opposition*, and is then at its closest to the Earth and shines at its brightest, being seen at its best in the middle of the night. When the angle between the Sun, Earth and planet is ninety degrees, the planet is said to be at *quadrature*. The period between successive oppositions of a planet is called its *synodic period* and is not the same as its orbital period

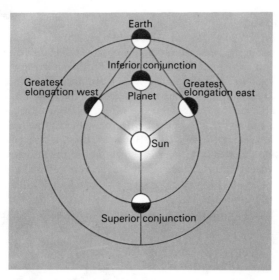

Conjunction and greatest elongation of a planet

simply because the Earth itself is moving and it is the relative motions of the two planets which give rise to such phenomena as oppositions.

When an inferior planet passes across the face of the Sun, it is said to be *in transit*. If the planes of the inner planets' orbits were exactly the same as the Earth's, then a transit would occur once every synodic period. However, this is not the case. Transits of Mercury are relatively rare (the next two will occur in 1970 and 1973) and those of Venus very scarce (next transit in 2004).

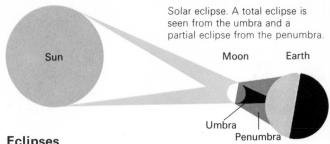

Solar eclipse. A total eclipse is seen from the umbra and a partial eclipse from the penumbra.

Sun

Moon    Earth

Umbra

Penumbra

## Eclipses

A *total eclipse* of the Sun occurs when the Moon passes exactly in front of the Sun and completely obscures the Sun's disc from view. It is then possible to see the Sun's outer regions, the *chromosphere* and the *corona*, which are normally too faint to see. Total eclipses are rare events because the *umbra*, the cone of dark shadow caused by the Moon, only just reaches the Earth and so the area of the Earth from which a total eclipse is visible is very small indeed. However, there is a large region outside the area of totality in which part of the Sun is seen to be obscured, the *penumbra,* and observers can see a *partial*

A total solar eclipse

*eclipse*. Partial eclipses are quite common, and there are at least two a year. The reason why an eclipse of the Sun does not occur every month is that the plane of the Moon's orbit is not quite the same as that of the Earth's and thus the Moon is more likely to pass above or below the Sun than actually in front of it.

It is a curious accident of nature that we can ever see exact total eclipses of the Sun at all. The Moon happens to be four hundred times nearer to us than the Sun, so that although the Sun has a diameter four hundred times greater than the Moon, they appear exactly the same size in the sky.

Eclipses of the Moon occur when the Earth lies between the Sun and the Moon (that is, at Full Moon) and the Moon happens to pass through the Earth's shadow. Since the Earth is four times larger than the Moon, it follows that the Earth's cone of shadow is correspondingly greater and eclipses of the Moon, total eclipses in particular, are more common than solar eclipses. During an eclipse of the Moon, the Earth's shadow on the Moon is clearly seen to consist of two parts: the dark inner umbra, where the Sun is completely cut off and the penumbra, where only part of the Sun is obscured.

A lunar eclipse

# ASTRONOMICAL INSTRUMENTS

The earliest astronomers had only their eyes with which to make observations and so observations tended to be rather inaccurate. However, such instruments as the sundial have been in use for many thousands of years—the ancient Chinese certainly knew of them. The simplest form of sundial is a post stuck in the ground: the approximate solar time is given by the position of the shadow of the post, if one knows that the Sun is due south (in the northern hemisphere) at midday. More sophisticated sundials can be calibrated quite accurately. As the Earth rotates from west to east, it follows that at any instant the time shown on a sundial in, say, England will be different from the time shown by a similar sundial in the United States. In fact, there is a difference of one hour for every fifteen degrees of longitude on the Earth. Thus if it were noon in England, it would be 7 a.m. in the eastern USA. The situation is further complicated by the fact that the Earth actually rotates on its axis in 23 hours 56 minutes, although, because of its motion around the Sun, the period between successive solar middays is 24 hours. *Sidereal time* is related to the actual rotation period of the Earth.

The medieval astronomers perfected several useful instruments such as the *astrolabe,* which was used for finding the positions of stars. Such instruments were painstakingly engraved by craftsmen in a most ornate fashion and remaining examples are highly valued. The *armillary sphere* was a sort of celestial globe showing such features as the zodiac and the celestial equator. Used for accurate mapping of stellar and planetary positions was the *quadrant*, with which the altitudes of stars above the horizon could be read. Tycho Brahe in the sixteenth century made measurements accurate to within a couple of minutes of arc using refined quadrants.

Astronomy was revolutionized by the invention of the telescope at the beginning of the seventeenth century by the Dutch spectacle maker Hans Lippershey. Lippershey's instrument used lenses, and this type of telescope is called the *refractor*. Certain difficulties with refracting telescopes led to the design by Newton and Gregory of telescopes using mirrors, this type being called the *reflector*.

Two of Galileo's telescopes

A sundial

(*Above*) An armillary sphere. The rings represent the principal circles of the celestial sphere. These devices were used in various forms by early astronomers to determine the positions of stars.
(*Left*) A quadrant

The Yerkes refractor at the Yerkes Observatory near Chicago. The focal length is 63 feet.

## The refractor

The principle of the refracting telescope is shown in the illustration. When a beam of light passes from a less dense medium such as air, to a denser one such as glass, it is bent, that is, refracted, from its original direction. A convex lens (known as a *positive lens*) will cause rays of light from a distant object to converge towards a point, called the *focus* of the lens, where a small image of the object will be formed. The distance from the lens to the focus is called its *focal length*, and the ratio of the focal length to the diameter, or *aperture*, of a lens is called its *focal ratio*.

Now, the refractor consists basically of two lenses, the *object glass* and the *eyepiece*. The object glass is a large aperture lens with a long focal length and large focal ratio which forms an image of the object in view. This image is then viewed and magnified by means of the eyepiece, which is a small, short-focal-length lens. The normal astronomical telescope uses a convex eyepiece and produces an inverted image. Galileo's

Object glass       Eyepiece

The principle of the refracting telescope

telescopes in fact used concave (negative) eyepieces which gave erect images. This system is used in opera glasses but is not very efficient. Normal terrestrial telescopes employ extra lenses to re-erect the image.

The simple refractor suffers from one very serious defect. White light is composed of a mixture of all colours from red to violet, and the trouble is that all these colours are refracted by differing amounts in the same lens. The result is that the focal length of a particular lens is much less for blue light, for example, than its focal length for red light, and so any object seen with the eyepiece will have bright coloured fringes around it. One way to minimize this effect is to make the object-glass focal length exceedingly long; this led to the construction of very unwieldy instruments. Chester More Hall virtually solved the problem when in 1729 he invented the *achromatic lens,* a compound lens made of different types of glass which tend to cancel out the false colour effect. The largest refractor in existence is the 40-inch-aperture giant at Yerkes Observatory near Chicago in the USA.

Chromatic aberration in a simple lens

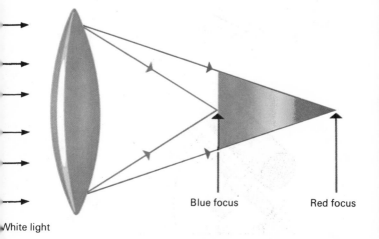

Blue focus      Red focus

White light

The principle of the Newtonian reflecting telescope

The principle of the Cassegrain reflecting telescope

Polar axis

Declination axis

The equatorial telescope mounting

## The reflector

The reflecting telescope replaces the object glass by a mirror. A suitably shaped concave mirror will reflect light rays back towards a focus and the resultant image can then be viewed by means of an eyepiece. In order to reflect all rays to the focus, the shape of the mirror surface has to be a paraboloid, but this is only slightly different from a spheroid which is easy to construct. In fact for small aperture mirrors, a spherical surface is good enough. Reflectors do not give any chromatic aberration.

There are various types of reflector, and the principles of two of these are shown here. One difficulty is where to put the eyepiece so as to see the image. Obviously, since the image is formed in front of the mirror, if the eyepiece were placed at the focus, the observer's head would block out all the light from the object being examined. In the Newtonian reflector, a small flat diagonal mirror is placed just before the focus. This reflects the focal point to the side of the telescope tube where the eyepiece is placed. In the Cassegrain system, a small convex mirror reflects the focal point back through a hole in the main mirror. This system has the advantage that a long focal length can be accommodated in a relatively short tube.

The largest reflecting telescope is the 200 inch Hale reflector at Mount Palomar, California.

## Telescope mountings

It is essential that powerful telescopes be rigidly mounted and that the mounting be such that the instrument can be easily directed towards any part of the sky and the apparent motion of the planet in the sky followed. The simplest mounting, the *altazimuth,* consists basically of a vertical and horizontal axis. Movement of the telescope round the vertical axis turns the telescope in a direction parallel to the horizon, whilst motion about the horizontal axis moves the telescope in altitude. If the vertical axis is tilted parallel to the axis of the Earth, then motion round this axis alone, the *polar axis*, will keep a planet in view. This type is called the equatorial mounting. All large telescopes are equatorially mounted and the polar axis driven at a suitable rate to compensate for the Earth's rotation.

## Specialized telescopes

Since the mid-nineteenth century, astronomical photography has become increasingly important and nowadays nearly all the work carried out at a large observatory is done photographically. Very little direct observation at the telescope is done and the popular picture of an astronomer sitting freezing all night at the eyepiece no longer applies. The use of photography has led to the development of specialized telescopes such as the Schmidt telescope.

Ordinary telescopes show only a very small area of sky – a degree or less – at a time and it is often desirable to photograph much larger areas than this. The *Schmidt telescope* ('camera' is perhaps a better description since it is never used visually) uses a short-focus spherical mirror to form an image of a large area of sky on a photographic plate placed at the focus. By itself the spherical mirror would distort the image, and so a carefully shaped glass plate is placed before the mirror to remove these distortions.

Photographic telescopes have to be very accurately driven since very long exposures of the photographic plate have often to be made in order to detect extremely faint objects.

(*Left*) A Schmidt telescope

(*Opposite*) The 250-foot radio telescope at Jodrell Bank, England

## Radio-telescopes

In 1933 an American radio engineer, Karl Jansky, investigating problems in radio interference, accidentally discovered radio waves reaching the Earth from space. This radiation was subsequently identified as coming from a distant source in the Galaxy. Since 1945 progress in radio astronomy has been rapid. Many cosmic radio sources are now known: the Sun, for example, emits radio waves, and there seems to be some sort of radio emission from the planet Jupiter.

There are many types of radio-telescope designed to collect these waves, from simple aerials to complex arrays of aerials. The type most analogous to optical telescopes is the dish-type reflector, of which the best-known example is the 250-foot-

A radio interferometer

diameter instrument at Jodrell Bank in England. Such instruments reflect radio waves to a receiver at the focus of the dish. The signal is then amplified and recorded.

Of the other types of radio-telescope, one of the most important is the *interferometer*, which makes use of the phenomenon of *interference*. Light and radio waves are examples of electromagnetic radiation, and both travel at 'the speed of light', 186,000 miles per second (300,000 kilometres per second). Electromagnetic waves can be regarded as being analogous to water waves, the distance from crest to crest being the wavelength, and these waves come in all lengths ranging from gamma and X-rays with wavelengths of about a thousand millionth of a centimetre to the sort of waves used for radio broadcasting with wavelengths of a kilometre or more. Light waves are of the order of a ten thousandth of a centimetre.

When two electromagnetic waves meet, they are said to interfere. If two crests happen to superimpose, then a stronger signal will result, whereas if a crest and a trough meet, no signal results. The radio interferometer consists of two or more receivers, dishes or otherwise, set a known distance apart. The signals they receive are combined, and by adjusting the angles of the receivers to get a certain type of interference pattern in the resultant signal, the position of a radio source can be accurately measured.

Long and short light waves

In radar astronomy, the distance of planets is measured by bouncing a signal off the planet and noting how long it takes for the signal to go out to the planet and return.

## The spectroscope

Newton showed that when white light is passed through a glass prism, it is split up into all the colours from red to violet. White light consists of a mixture of light of differing wavelengths, each wavelength corresponding to a certain colour. The shorter the wavelength of light, the more it is refracted in the prism, and thus blue light which has a short wavelength is refracted more than red light which has a longer wavelength. When light is spread out in this way, it is called a *spectrum*. The spectra of the Sun, stars and planets contain characteristic patterns of dark lines caused by the absorbing effects of the surrounding layers of gases of various elements. The spectroscope shows these patterns and enables astronomers to discover what materials are present in distant bodies.

The principle of the spectroscope

Jupiter as seen through a large telescope

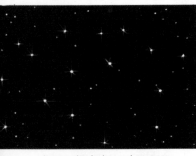

Atmospheric haze decreases greatly the number of stars we can see. (*Above*) Part of the sky as it would appear if there were no atmosphere and (*below*) stars seen through atmospheric haze.

## Resolving power

An important factor in any telescope used for exploring the planets is its *resolving power*. In other words, what size is the smallest feature which it can distinguish on a planetary surface? Resolving power is, in fact, defined as being the angular separation of two point sources of light which can just be seen to be separated in the field of view of a telescope, and this gives a good guide as to the smallest feature which can be seen on a planet. The resolving power $R$ of a telescope is given in seconds of arc by the simple formula $R = \dfrac{4 \cdot 56}{d}$, where $d$ is the aperture of the telescope in inches. Thus a 2-inch refractor can show features down to about 2·3 seconds, and a 10-inch reflector down to 0·46 seconds. Observing the Moon, such resolving powers would correspond to craters of about $2\frac{1}{2}$ miles and $\frac{1}{2}$ mile in diameter respectively.

To calculate the size in miles of the smallest feature which a given telescope can show on a planet, it is neces- to know the angular diameter of the planet. The angular diameters of the planets depend on their distance from the

Earth and their size. Superior planets will be at their nearest and will have greatest angular diameter at opposition, whilst inferior planets will appear largest at inferior conjunction. The maximum angular diameter of Jupiter is about 50 seconds, while even at the most favourable opposition, the maximum Mars can attain is 25 seconds. Considering Mars at a good opposition, a 10-inch telescope will show features about 0·5 seconds across, or about a fiftieth part of the planet's diameter. Mars is roughly 4,000 miles in diameter, and so the smallest feature this telescope should show would be about 80 miles across.

Most nights, even when the sky is clear, a telescope will not resolve down to its theoretical limit because of the effect of the atmosphere on light from the planets. Often a planet appears to be shimmering and unsteady (astronomers say the 'seeing' is bad) and only occasionally is the image crystal-clear.

The Lick Observatory, California. The large dome houses the 120-inch reflector.

| γ-rays | X-rays | Ultra-violet | Infra-red | Radio | Waves |

Optical window / Radio window

Absorption      Absorption      Absorption

$10^{-8}$ cm    $10^{-6}$ cm    $4 \times 10^{-5}$ cm    $8 \times 10^{-5}$ cm    $10^{-1}$ cm    $10^{2}$ cm

The optical and radio windows

## The problem of the atmosphere

Because of the atmosphere, the only radiations which reach the Earth largely unimpeded are visible light and certain radio waves in the wavelength range 0·1 centimetre to 100 centimetres. These small parts of the electromagnetic spectrum are known as the *optical window* and the *radio window* respectively.

As our eyes can only see visible light, that is, that narrow part of the electromagnetic spectrum with wavelengths bet-

Part of the Earth's surface as seen from a satellite

ween 4,000 and 7,000 Angstrom units (one Angstrom unit is a hundred millionth of a centimetre), we need other forms of detectors to record other radiations

X-rays and ultra-violet light can be detected photographically using suitable film, but no useful attempt to do this can be made from observatories on the ground. However, some radiation of longer wavelength than visible light and shorter than radio waves, can get through. These waves are the *infra-red,* and it is this part of the spectrum that carries heat. Suitable detectors have been constructed to record this radiation, but at ground level results are poor. Water vapour and carbon dioxide gas are particularly efficient at stopping infra-red rays, and the atmosphere contains large quantities of these.

The only real answer to the problem of the atmosphere is to get above it. A space observatory would be able to observe the whole electromagnetic spectrum twenty-four hours a day. Observing the planets in the ultra-violet and infra-red would be a great boon. A 15-inch optical telescope would be as effective as the largest telescope at ground level.

X-ray picture of the Sun taken from a rocket (Official U.S. Navy Photograph)

## SPACECRAFT

For countless years astronomers have dreamed of getting above the atmosphere, and indeed for some decades now it has been possible to lift astronomical instruments above a large proportion of the atmosphere by means of balloons filled with hydrogen or helium gas. Helium is used most commonly, as hydrogen forms an inflammable mixture with oxygen. Since these gases are much lighter (less dense) than air, it follows that balloons so filled will tend to rise above the dense lower parts of the atmosphere until they reach a level where the density of the air is much the same as that of the gas in the balloon. Thus small quantities of equipment can be raised to heights of over twenty miles. But this is hardly space flight.

In the second century AD the Greek writer Lucian wrote one of the earliest science fiction stories, in which a ship full

Launching a high-altitude balloon

(*Left*) The padded cabin of
Jules Verne's space projectile
(*Above*) Robert Goddard
launched the world's first
liquid-fuelled rocket in 1926.

of men was picked up by a waterspout and hurled up to the
Moon. Many other such stories followed, including one in
which the hero was towed to the Moon by a flock of wild
swans. A rather better idea was suggested by the famous
French author Jules Verne in his book *From the Earth to the
Moon*. His idea was to fire a large shell containing his space
travellers from an enormous gun with a speed of 7 miles per
second, sufficient to take it to the Moon. Unfortunately such
a shell would probably be destroyed by atmospheric friction,
and in any case the shock of such sudden acceleration would
certainly kill the crew members.

Nevertheless Verne was correct in one respect: a projectile
given a velocity of 7 miles per second could certainly reach
the Moon if it were aimed correctly. This speed is the Earth's
escape velocity. If you throw a stone upwards, it will reach
a certain height and then fall back. If, however, you could
find a means to throw the stone at the escape velocity, it
would completely overcome the Earth's gravitational pull
and continue to travel outwards forever.

(*Above*) The launching of a V2 rocket

(*Below*) The principle of the rocket motor. In *a* the nozzle is closed and internal pressure forces are in equilibrium. In *b* the nozzle is open and the force acting on the opposite face is no longer balanced. The result is motion.

## Rockets

Aircraft would be useless in space for two reasons. Firstly their wings require air to provide lift to keep them flying, and secondly propellers require air to work on to provide forward thrust. Even the jet engine requires a plentiful supply of air and there is of course no air in space. Thus travel to the planets requires a totally different method of propulsion – the rocket. All rockets, from fireworks up to interplanetary craft, utilize what Newton termed the *principle of reaction*. This principle is easily illustrated by means of a balloon. When a balloon is fully inflated and the end sealed, there is an equal pressure of gas at every point within it. If the nozzle is now opened, gas will escape rapidly through the nozzle and the balloon will shoot off in the opposite direction.

A balloon so released does not get very far – it soon uses up all the gas it contained and is slowed down by friction with the atmosphere. The rocket motor carries its fuel with it, and this is burned inside the motor to produce hot gases at a controlled rate. The expanding gases provide the thrust that propels the rocket. Many smaller rockets (including fireworks) use solid fuel packed into the motor, but most larger ones use liquid fuels pumped from tanks into the motor at a suitable rate.

The theory of the liquid-fuelled rocket was published as long ago as 1903 by the Russian Tsiolkovsii, but his ideas were neglected for several decades. Independently, in the USA, R. H. Goddard published a book showing the possibilities of rocket propulsion, including the idea of firing one to the Moon. In 1926 he fired the first liquid-fuelled rocket, which, using a mixture of petrol and liquid oxygen, reached a height of about 200 feet. Another great pioneer was Oberth, whose publications led to a great interest in rocketry in Germany. During the Second World War the V2 rocket was perfected, which could attain a height of 100 miles and which was used to drop explosives on London in 1944–5. After the war, V2 rockets were used to explore the upper atmosphere and some of the inaccessible parts of the electromagnetic spectrum.

The V2 was a highly successful rocket and can certainly be regarded as the true forerunner of modern spacecraft. Improvements were rapid and a great deal of research was done

Large modern rockets consist of several stages which ignite in turn.

with such rockets as the Skylark and Viking. However, the heights that such rockets could attain were distinctly limited. The problem is that known rocket fuels can only provide a limited amount of energy, and the heavier a rocket is, the less it can be accelerated by the amount of thrust available from its motors. Ideally a rocket should be as light as possible, but it has to be rigidly constructed to withstand tremendous stresses and strains, and it has to carry the weight of its own motors and fuel tanks, the result being that no simple rocket can reach anything like escape velocity.

This problem has been overcome by means of the step-rocket, which consists of two or more rockets placed on top of each other. When the bottom rocket, or first stage, has used up all its fuel, it drops off, leaving the second and subsequent stages to fire in succession, each starting off with the benefit of the previous stage's velocity. In this way, a large amount of useless weight is ejected, and the final stage can be made to reach the desired speed with its payload of scientific equipment. The first such step rocket was fired in 1949, and consisted of a small rocket, the WAC Corporal, placed on top of a V2. The final altitude attained was 250 miles.

A great deal of research has been done with rockets of all sizes, including the ingenious little rockoons (small rockets fired from high altitude balloons), into the properties of the upper atmosphere. We owe all our knowledge of the way temperature varies with height, for example, to the rocket. The ionosphere (the region of the atmosphere which reflects radio waves) has been studied, together with ultra-violet, X-ray and other wavelengths from the Sun and outer space. However, the big drawback with such experiments is the limited amount of time—a matter of a few minutes—that the rocket remains at a suitable height. Often, too, instruments and records are damaged on impact when they return.

## Satellites

Lengthy observations above the atmosphere can only be carried out by means of artificial satellites, that is, equipment placed in orbit around the Earth, and plans were announced by the USA in 1955 to do just this. As we said earlier, the Moon is only prevented from falling into the

Earth by the fact that it is moving fast enough to counteract the force of attraction. The critical velocity, parallel to the Earth's surface, which a satellite must attain to stay in orbit round the Earth is called orbital velocity and is roughly 5 miles per second, just under 18,000 miles per hour for an orbit 000 miles above the ground. The orbital velocity decreases the further away one moves from the Earth: at the distance of the Moon a satellite would only need to move at about 2,000 miles per hour, the speed at which the Moon moves. Once in orbit above the atmosphere, a satellite will continue in that orbit simply because there is nothing to stop it.

The Americans planned to launch their first satellite as part of the programme for the International Geophysical Year of 1957–8 but, much to everyone's surprise, the first-ever artificial satellite was launched by Russia on 4 October 1957. This was Sputnik 1, a sphere 23 inches in diameter weighing 184 pounds, orbiting in an elliptical orbit with an apogee of

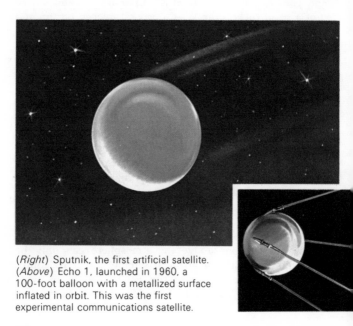

(*Right*) Sputnik, the first artificial satellite.
(*Above*) Echo 1, launched in 1960, a 100-foot balloon with a metallized surface inflated in orbit. This was the first experimental communications satellite.

588 miles and a perigee of 142 miles. The orbital period was 96 minutes. At the perigee distance there is still an appreciable quantity of atmosphere and this produced a frictional drag on the satellite, so that the orbit slowly changed until finally, on 4 January, 1958, Sputnik 1 dropped into the denser layers of the atmosphere and was burned up by friction, just like a meteor. Sputnik 1 was followed by Sputnik 2, and on January 31, 1958 the Americans placed in orbit the tiny 31-pound Explorer 1, which was however packed with instruments.

One of the most spectacular discoveries by the early Explorers 1 and 3 was the hitherto unsuspected belts of charged particles surrounding the Earth at a height upwards of 600 miles. These were named the Van Allen belts. Subsequent Explorers and Pioneers enabled scientists to map these belts and showed that they extend to an altitude of at least 25,000 miles.

Perhaps the best known, certainly the most widely observed, early satellite was Echo 1 launched in 1960.

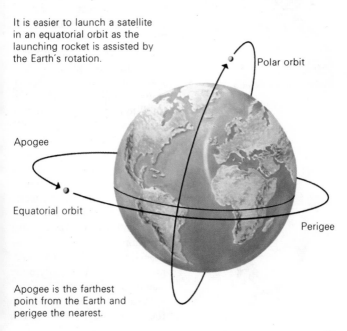

It is easier to launch a satellite in an equatorial orbit as the launching rocket is assisted by the Earth's rotation.

Polar orbit

Apogee

Equatorial orbit

Perigee

Apogee is the farthest point from the Earth and perigee the nearest.

American satellite Tiros

British satellite Ariel 3

Russian satellite Lunik 3

This was a giant balloon satellite, inflated in space to a diameter of about 100 feet, and covered with reflecting material. It was used as a passive communications satellite; that is, radio waves were bounced back to the Earth from its surface. It was readily visible to the naked eye, looking like a bright star moving fairly rapidly across the sky, and continued in orbit for several years.

The next stage was active communications satellites containing relaying equipment. The American Telstar successfully relayed part of the 1964 Olympic Games from Tokyo, but because of its orbital speed it appeared to cross the sky quite quickly and could only carry short transmissions at a time. If a satellite is placed in orbit at a height of 22,000 miles, it will orbit the Earth in exactly one day and will thus appear stationary in the sky, allowing twenty-four-hour communication. The Syncom series of satellites did this. The American COMSAT corporation operates such satellites commercially and relayed most of the 1968 Olympic Games from Mexico.

Weather satellites such as the Tiros and Nimbus series

Clouds over Burma as seen from a spacecraft

have proved to be of great value in plotting cloud patterns on the Earth and giving early warnings of hurricanes and so on.

Several British and French satellites have been launched by American rockets and France, Japan and China have developed their own launchers.

## Probes to the Moon and beyond

At an early stage attempts were made to send space probes to the Moon, the first successful attempt being the Russian Lunik 1, which passed within 4,000 miles of the Moon in January 1959. Lunik 2 hit the Moon in September of that year, and in October Lunik 3 passed round the far side of the Moon, sending back pictures of the side which had hitherto never been seen.

Subsequent Russian and American spacecraft have dis-

Mariner space probe

covered a tremendous amount about the Moon, as we shall see later (page 78). The accuracy required in a Moon shot is quite fantastic, but when it comes to launching spacecraft to the planets the problems are magnified many times. Even at its closest, Venus is still more than a hundred times as far away as the Moon, and Mars half as far again. Also, the amount of power needed to fire a rocket directly, that is, in a straight path, towards Venus or Mars is far more than present-day motors can sustain, so that long, curving orbits, taking several months, have to be followed to these planets and only at infrequent intervals are they in suitable positions for such launchings.

For a voyage to Venus a spacecraft would be launched at dusk against the direction of the Earth's motion with a speed rather greater than escape velocity. Thus the spacecraft's velocity with respect to the Sun would be less than that of the Earth and the spacecraft would tend to fall in towards the Sun, intersecting the orbit of Venus on the way. The first successful space probe of this nature, Mariner 2, passed close to Venus in 1962 after a journey of 109 days.

For a trip to Mars the spacecraft would be launched at dawn in the same direction as the Earth's motion and would thus be moving faster than the Earth. The space probe would then move out from the Sun to intersect the orbit of Mars. In 1965 Mariner 4 successfully flew by Mars, sending back a series of spectacular photographs.

Once a spacecraft has been launched by a step rocket, it is seldom found to be on its precise planned course, simply because so many factors can produce minute deviations from the computed orbit. Thus space probes are fitted with small control jets, little rocket motors with sufficient fuel to make at least one major change in course during the flight. These controlling devices are operated from the Earth by radio. Information is likewise sent back by radio. Virtually direct television pictures can be sent from a spacecraft on its way to the Moon, but the planets are so far away that pictures taken by the spacecraft are sent back by radio suitably coded.

*(Left)* Mariner 2's orbit to Venus. *(Right)* The use of control jets. Small auxiliary rockets (often gas cylinders) can cancel out spin and turn the rocket about its centre of gravity (C.G.).

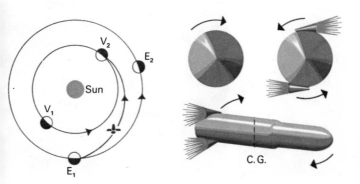

## Manned space exploration

What arouses most popular interest in the exploration of the planets is certainly the prospect of manned voyages through space. The first living creature to be sent into space in a satellite was the dog Laika in the Russian Sputnik 2 of November 1957. Before a manned flight could be attempted, it was essential to find out what effects space travel might have on living creatures. Would the level of radiation, for example, be harmful to man? When a satellite moves in an orbit round the Earth, the occupants experience weightlessness since the gravitational attraction of the Earth is balanced out by the motion of the satellite. What sort of effect might this have? It was for reasons such as these that Laika and later animals were sent into orbit. Unfortunately Laika never returned to Earth.

In the intervening years several unmanned spacecrafts were sent into orbit and then brought back to Earth, and finally, on 12 April 1961, the Russian Major Yuri Gagarin became the first man in space in his $4\frac{3}{4}$-ton spacecraft Vostok 1. He made a single orbit of the Earth at heights ranging from 112 to 203 miles in a period of 1 hour 48 minutes and returned safely to a prearranged spot in the Soviet Union. By his flight he had shown that man could definitely withstand the accelerations involved, weightlessness, radiation and all the other postulated hazards, at least for short periods, and he was acclaimed throughout the world. Tragi-

(*Left*) Yuri Gagarin, first man in space. (*Right*) Vostok spacecraft

Edward White's space walk during the Gemini 4 flight

cally, he was killed in a plane crash in 1968. On 5 May 1961 Commander Alan Shepard became the first American in space with a fifteen minute up and down flight to a height of 120 miles.

The next flight in the Vostok series was more ambitious. Major Titov in Vostok 2 made seventeen and a half complete orbits on 8 and 9 August 1961, returning without any ill effects. The American manned programme came under the title of project Mercury and their first orbital flight was made in February 1962 by Colonel John Glenn, who in his craft Friendship 7 made three orbits.

Now, only a few years later, manned orbital flights are almost commonplace. Man has landed on the Moon, and contemplates flight to Mars.

Mercury and Venus in the evening sky. Mercury is just visible, white near the horizon. Venus is well above the horizon, bright and silvery.

Map of Mercury based closely on that of E. M. Antoniadi

# THE INNER PLANETS

The two inferior planets are Mercury and Venus, the messenger of the gods and the goddess of love. As they lie closer to the Sun than the Earth does, they tend to lie quite close to the Sun in the sky and so when they are visible it is usually just after sunset or just before sunrise. Mercury is particularly elusive, and it is said that Copernicus never saw it in his life. Venus, on the other hand, is quite unmistakable when well placed. It is then far brighter than anything else in the sky apart from the Sun and Moon, and can even on occasion cast a shadow. Let us see what astronomers have been able to find out about these planets with their telescopes and space probes.

## Mercury

Mercury, with a diameter of about 2,900 miles, is the smallest planet in the solar system (although it is just possible that Pluto is smaller). It is also the closest to the Sun, moving at an average distance of 36 million miles with an orbital period of 88 days. In fact Mercury's orbit is highly eccentric, so that the planet approaches within 29 million miles of the Sun at perihelion and moves out to a distance of 43 million miles at aphelion. Because of this eccentricity, the planet's angular distance from the Sun at greatest elongation can vary from 18 to 27 degrees, and thus all apparitions of Mercury are not equally favourable.

The early observers found it rather hard to see any detail on Mercury at all because of its small size and the fact that when it was visible in the night sky it was near the horizon and so shining through a large amount of our unsteady atmosphere. The phase could be made out, but that was about all. Schroter, working in the late eighteenth and early nineteenth century, tried to draw a map, without much success, and the first reasonable map of the surface markings was drawn some decades later by the Italian Schiaparelli. E. M. Antoniadi, working in France, produced a map in 1933 as a result of his work with a 33-inch refractor. His chart remains the best available, although still very approximate.

From his observations of the surface features he had mapped, Schiaparelli concluded that Mercury kept the same

Artist's conception of the surface of Mercury

face turned towards the Sun all the time (just as the Moon does to the Earth) and thus the planet's rotational period is the same as its 'year', namely 88 days. This conclusion was confirmed by the few observers who bothered to study the planet in later years. With its small size, Mercury has been unable to retain any appreciable atmosphere. This is because the gas molecules composing any atmosphere Mercury might once have had would have been moving about at speeds faster than the planet's escape velocity, and so would have escaped completely from the planet. Thus there can be no warm air currents travelling from the sunward side to the dark side, and so the hemisphere of the planet facing the Sun should reach temperatures of 400°C, while on the dark side the temperature should fall well below —250°C.

The 88-day rotation period was accepted without question for about a hundred years, despite the fact that early in this century thermocouple measurements of the infra-red radiation from the dark side of Mercury indicated a temperature much higher than would be expected if the dark side never saw the Sun. The results were regarded as rather unreliable. In the past decade even more puzzling temperature estimates were made, including values of $-23°C$ and even $+27°C$ for the dark side, from radio and microwave methods. Part of the discrepancy can be explained by the fact that different methods of measurement indicate temperatures at different levels below the planet's surface; however, all the results indicate that Mercury cannot keep the same face toward the Sun all the time.

In 1965 the 1000-foot radio dish at Arecibo in Puerto Rico was used to determine the rotation period by a radar technique If a planet is rotating on its axis, then as we look at it, one side of the planet will be approaching us and the other receding. If now we transmit a radio beam towards the planet, then the part of the beam which is reflected from the side approaching us will return with its wavelength somewhat reduced while the part reflected from the receding side will return with a slightly longer wavelength. This is an example of the Doppler effect, and the returning signal will show a certain characteristic pattern from which the rotation rate can be found. The Arecibo results showed the rotation period to be 59 days, and this value has since been well confirmed.

The interesting thing about the 59-day rotation period is that it is almost exactly two-thirds of the planet's period of revolution around the Sun, and half the synodic period. Thus it happens that when Mercury returns to a position in the sky at which it has previously been observed it has rotated twice on its axis, and the same features will be visible in the same positions once more. This explains why the early observers were fooled into thinking that Mercury had a captured rotation.

Because of the difficulties in observing the planet from the Earth, many problems concerning Mercury may not be resolved until a space probe can be sent to its vicinity. A spacecraft launched in October or November 1973 could be aimed in such a way that Venus could be used to deflect it to Mercury. The United States intends to attempt this, and the European Space Research Organization has similar plans.

## Venus

Venus, with a diameter of just over 7,700 miles, is very nearly the same size as the Earth and has often been described as the Earth's twin, although—as we shall see later—size is about the only thing the two planets have in common. The planet is completely swathed in dense, highly reflective clouds and so appears very bright in the sky. The amount of light which a planet reflects is called its *albedo*—the maximum possible albedo, when all the incident light is reflected away again, is 1, and the minimum, when no light is reflected, is 0. On this scale, Venus has an albedo of 0·59. Mercury, on the other hand, has an albedo of only 0·06, similar to that of the Moon.

When Venus is visible at an evening elongation, it is often known as the *evening star*, whilst at a morning elongation, it is the *morning star*. The ancient astronomers thought for a long time that the morning and evening stars were different bodies, calling them Phosphorus and Hesperus, but Pythagoras realized in 500 BC that the two must be the same. Because of its brightness and relatively rapid motion, it is quite possible that Venus was the first planet to be recognized as such.

Venus, like Mercury, goes through a complete sequence of phases in each synodic period. Starting from full at superior

Earth and Venus compared. Venus' dense dusty cloud layer completely obscures the planet's surface.

conjunction, the illuminated phase dwindles to half at greatest elongation, after which it becomes an increasingly narrower crescent until inferior conjuction when it is invisible. The process is then repeated in reverse until the next superior conjunction. Because the distance of Venus varies, its angular diameter ranges from a mere $9\frac{1}{2}$ seconds at superior conjunction to 65 seconds at inferior conjunction. The angular diameter at half phase, or dichotomy, is about 25 seconds. Because of the variation in phase and distance of Venus, the apparent brightness varies in a complicated way, and in fact the planet appears brightest during the crescent phase when 30 per cent of the illuminated side is visible. The angular diameter then is about 35 seconds. It may seem puzzling that greatest brilliancy occurs during the crescent phase, but increasing phase has to be balanced against decreasing apparent diameter.

The inner planets appear at their largest but dullest when they are at inferior conjunction, that is, when they are closest to Earth. They appear brightest in the crescent phase.

S

Lowell's 'canals' on Venus

The early telescopic observers found Venus rather disappointing. Apart from the phase, hardly anything could be seen on the face of the planet except a few very vague and ill-defined shadings. Bianchini, working in the early eighteenth century, claimed to see quite well-defined markings and went so far as to produce a map showing seas and continents, but the telescope he used was decidedly clumsy and non-achromatic and later observers were unable to confirm these markings. In the twentieth century, some very interesting observations were made in Arizona by Percival Lowell. Lowell recorded well-defined dusky radial streaks on the disc of Venus, and drew a map showing these features, which he was convinced were surface features.

Other observers with better equipment were unable to see any of Lowell's features, and it is now accepted that the markings which Lowell, or Bianchini for that matter, saw were wholly illusory. Observing such a bright disc as Venus tends to produce optical illusions such as the spoke-like features which Lowell recorded.

As it is now well established that Venus is covered in a dense cloudy mantle, what can be seen? Ill-defined dusky shadings can certainly be observed. These dusky patches must signify cloud patterns of some sort, and can be studied in greater detail through colour filters which only allow light

Vague shadings are apparent on the cloud mantle of Venus. A cusp cap can be seen clearly (*above*) at the southern (top) point of the planet's crescent.

of certain wavelengths to pass through. Often, near the cusps (the points or horns of the crescent) bright patches known as cusp caps can be seen, sometimes surrounded by dark collars. Some astronomers maintain that these caps are optical features caused by contrast but they do seem to have a basis of reality. Unsuccessful attempts have been made to link them to the position of Venus' poles.

Another feature of Venus, presumably due to the cloudy

atmosphere, is the phase anomaly, or Schroter effec., whereby at morning elongations Venus tends to reach dichotomy later than it ought, while at evening elongations dichotomy is usually early. The discrepancy is usually a matter of a few days, but in exceptional cases, dichotomy can be out by as much as a fortnight. Amateur observers have, however, shown that in most instances the observed phase is smaller than the predicted phase.

The ashen light of Venus

Looking at the crescent Moon at night, it is often possible to see the dark side faintly illuminated, a phenomenon often called 'the old Moon in the young Moon's arms'. This is caused by light reflected from the Earth falling on the dark side. In the case of Venus in the crescent phase, it is sometimes possible to see a similar effect, known as the *ashen light*. The cause is hard to imagine. Since Venus has no moon, reflected light cannot be the answer (the Earth is too far away to have any effect whatsoever). One suggestion, was that the ashen light was due to the lights of enormous cities on the planet, but this seems highly unlikely. What did seem a reasonable idea was the suggestion that the effect was due to auroral-type displays in the upper atmosphere of Venus (for more about aurorae, see page 70). This required that Venus have a magnetic field like the Earth's, which seemed quite likely. But the Mariner 2 space probe showed that Venus has no appreciable magnetic field, and this has been confirmed by later spacecraft. The ashen light thus remains a mystery.

A Russian Venus probe

Another question which has been debated for many years is the problem of the rotation period of Venus. The problem is that with optical telescopes the only method of estimating this was to attempt to relate any observed motions of the dusky patches to the planet's motion. Cassini in 1666 estimated 23 hours 21 minutes, whilst Bianchini proposed 24 days 8 hours. In 1877 however, Schiaparelli came to the conclusion that the Venusian (or Cytherean) day was the same as its orbital period, 225 days. In 1956, radio observations by J. D. Kraus suggested a period of 22 hours 17 minutes.

Space probes seem to indicate that Venus has localized hot and cold spots on its surface. The red lines are isotherms, lines connecting points of equal temperature. This is a hot area near the southern cusp of Venus, caused perhaps by internal activity.

(*Opposite page*) Cross-section through the atmosphere of Venus.

The whole question was thus completely open when Mariner 2 approached Venus in 1962. Results sent back gave a rotation period of 243 days, which has since been confirmed. Thus Venus takes longer to rotate on its axis than to revolve around the Sun, a very strange state of affairs. An observer on Venus would see the sun rise in the west and set in the east.

The temperatures prevailing on Venus have long been a problem, and in many ways this is linked to the composition of the planet's atmosphere. Since Venus lies nearer to the Sun than the Earth does, it was reasonable to assume that its average temperature was higher than that of the Earth, but not drastically so. On the other hand, it was hard to estimate what effect the clouds would have on the surface temperature without knowing their composition, thickness and density. In the early part of this century, measurements of the temperature of Venus were attempted using a thermo-couple on the 100-inch reflector at Mount Wilson in the USA. The thermo-couple consists basically of two wires of different metals joined together; if one of the joints is warmed (by the infra-red radiation from Venus, in this case) a small electric current flows, giving an indication of the amount of heat. By this means, temperatures of about —35°C were estimated. This may seem peculiarly cold, but these temperatures in fact refer to the top of the cloud layer, quite high in the atmosphere of the planet, and we would expect such temperatures to be very low. Our own atmosphere rapidly gets cold with increasing height.

This still gave no clear idea of the surface temperature. Mariner 2 in 1962 showed that temperatures from 130°C to 330°C seemed to prevail, with some anomalous hot and cold spots. These results were disputed at the time. In October 1967, both the American Mariner 5 and the Russian Venus 4 arrived in the vicinity of the planet—Mariner continued past Venus, but the Russian craft succeeded in landing safely, a truly tremendous feat. Venus 4 results suggested even higher temperatures than Mariner 2—someone suggested that it might have landed down a volcano—though Mariner 5's results did not quite agree. Reinterpretation of the results has brought them into closer agreement, and Venus is certainly an exceed-

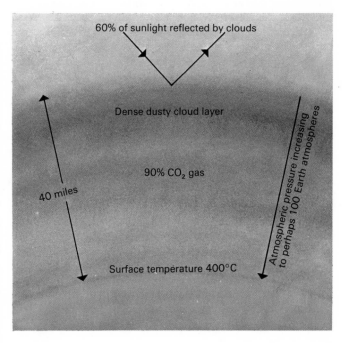

60% of sunlight reflected by clouds

Dense dusty cloud layer

90% $CO_2$ gas

40 miles

Surface temperature 400°C

Atmospheric pressure increasing to perhaps 100 Earth atmospheres

(*Left*) The Earth and Moon as they would be seen from Venus. The observer would of course have to be above the cloud layers.

(*Opposite page*) Early in this century it was proposed that Venus might bear life such as was seen on Earth millions of years ago.

ingly hot place. The atmosphere contains large quantities of nitrogen and carbon dioxide gas, with virtually no oxygen and water vapour. The clouds seem to be largely composed of dust. The atmosphere is extremely dense, at least fifteen to twenty times denser than our own, and the nature of the surface can only be speculated.

Many things remain uncertain about Venus, although space probe results have certainly revolutionized our ideas about the planet. The discrepancies between Russian and American results have been largely eliminated by closer analyses of the data, but a measure of confusion still remains. It is possible, for example, that the atmospheric pressure at the surface may even be as high as one hundred times that of the Earth's atmosphere. All that can definitely be said is that the atmospheric pressure is very high, and that temperatures there are quite unbearable, in excess of 350°C. Although the clouds reflect away a large proportion of the heat reaching Venus, much of the radiation which does get to the suface is trapped in the atmosphere by the effect of carbon dioxide gas and reflection from the underside of the clouds. It has even been suggested that, because of its density, the atmosphere will refract light rays all the way around the planet, and so an observer should, in theory, be able to see an image of the back of his head.

At one time, Venus was regarded as being a possible life-bearing planet. The Swede Arrhenius suggested in 1918 that beneath the clouds Venus was hot and humid—rather like our tropics taken to the extreme—and dense luxuriant plant life grew there at a tremendous rate. In fact, many people have held the view that Venus might be in a similar state to that of the Earth about 200 million years ago. Another suggestion was that Venus was covered in seas full of primitive aquatic organisms. Alas, these attractive ideas must be rejected: no form of life as we know it could exist in the conditions which are now thought to hold.

Any space traveller visiting Venus would need to go in a spacecraft which could withstand excessive pressures and temperatures. If he were to step outside, he would immediately be crushed and incinerated – not an attractive prospect! The view from above the clouds would be spectacular. As the Earth is fully illuminated from the point of view of Venus at closest approach, the Earth would be seen far brighter than Venus appears to us, looking a blue-green colour, with the yellow Moon in attendance.

(*Right*) Earth and Moon compared. The diameter of the Moon is just over a quarter that of the Earth.

(*Below*) The barycentre is the centre of mass of the Earth–Moon system and both bodies revolve around this point.

(*Bottom*) Tides are greatest when the pulls of Sun and Moon are combined as in *a*, least when Sun and Moon are at right angles as in *b*.

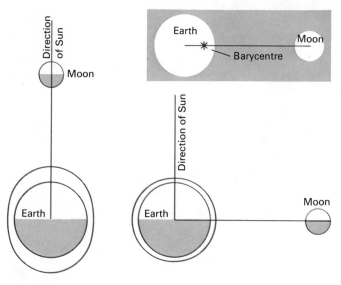

# THE EARTH-MOON SYSTEM

The system composed of the Earth and Moon is considered by many people to be more like a double planet than a planet with a satellite. Although most of the other planets have satellites (or moons if you like), these are always very small compared with the parent planet. The Moon, however, has a diameter of 2,160 miles compared to the Earth's 7,920, more than a quarter of the Earth's in fact. The volume of the Moon is about 1/50 of the Earth's and its mass is only 1/81 that of the Earth. However, this is still a far larger fraction of the parent planet's mass than is the case with any other satellite in the solar system.

The Moon revolves around the Earth at an average distance of 239,000 miles in a period of 27 days 7 hours 43 minutes, while the Earth revolves around the Sun in a period of $365\frac{1}{4}$ days. Because of this motion, the time taken by the Moon to complete its cycle of phases from new to new again is $29\frac{1}{2}$ days (its synodic period): the Moon has to travel further to get in line with the Sun again than it would if the Earth was stationary. It is not strictly true to say that the Moon revolves around the Earth. If we consider a barbell, this will balance at the centre of the rod joining the two weights—this point being called the centre of mass. The Earth-Moon system can be regarded as a barbell with one weight 81 times heavier than the other, so that the centre of mass lies near the heavier weight. Both Earth and Moon revolve around this point, the *barycentre*.

The main cause of tides on the Earth is the pull of the Moon, which tends to bunch up the water on the part of the Earth directly under the Moon. A lesser 'mound' of water is built up on the other side, each mound corresponding to a high tide. The Sun has a lesser effect; very large tides occur when Sun and Moon pull in the same direction, very small ones when they pull at right angles to each other.

## The Earth
The Earth moves round the Sun in an orbit of low eccentricity (0·017), at an average distance of 93 million miles rotating on its axis in a period (measured with respect to the stars)

of 23 hours 56 minutes. The Earth's axis is not perpendicular to the plane of its orbit. It differs from the perpendicular by $23\frac{1}{2}$ degrees. This gives rise to the seasons, as the northern winter and southern summer occur when the north pole points away from the Sun, while the northern summer and southern winter occur when the north pole points more towards the Sun. Southern summer occurs when the Earth is near perihelion and moving fastest, and it thus tends to be hotter and shorter than the northern one.

The average density of the Earth is about 5·5 times that of water. The density varies from very high values (10 to 12) at the centre of the globe to about 2·5 to 3·5 for the surface rocks. The internal structure of the Earth has been investigated by observing the effect that passage through the Earth has on shock waves from explosions and earthquakes. The Earth has a dense metallic core of radius 2,100 miles, surrounding which is a layer of heavy rock some 1,800 miles thick decreasing in density outwards. On top of this lies the lighter crust, less than 50 miles thick. Life depends upon the few feet of soil on top of this, the oceans and atmosphere.

The atmosphere is composed mainly of two gases, nitrogen (78%) and oxygen (20%), oxygen of course being vital for ou·

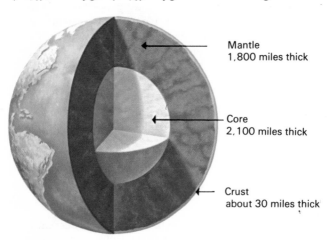

Mantle
1,800 miles thick

Core
2,100 miles thick

Crust
about 30 miles thick

A section through the Earth

form of life. Other gases make up the rest, carbon dioxide being one of the most important. The atmosphere prevents harmful radiations reaching the ground, and carbon dioxide, water vapour and clouds prevent all the heat which reaches the ground from being reflected away again. Thus clear nights tend to be colder than cloudy ones. There are traces of atmosphere hundreds of miles up, but most of the mass of air is concentrated in the first five to ten miles.

The Earth possesses a magnetic field—this is what causes compass needles to point north—and this seems to be linked in some way to the metallic core of the planet. However, the reason for the Earth's magnetism is not wholly understood, nor is the reason why the north magnetic pole moves. The

A section through the atmosphere

Height in miles

**Ionosphere** — 80

Meteors and aurorae — 60

— 40

**Stratosphere** — 20

Cirrus and cumulus cloud — 10

**Troposphere** Sea level

Ocean deeps 36,000 feet    Everest 29,000 feet

Earth's field attracts towards the poles charged particles emitted from the Sun and it is these particles which cause the aurorae. The aurorae are usually only seen from high north or south latitudes. They appear as moving bands, rays and arcs of light, usually of greenish or pink hue. Auroral displays often cause severe radio interference because of the effect that charged particles have on the *ionosphere*, the layer of the atmosphere about 75 miles up which reflects radio waves round the Earth.

Most people have seen meteors or 'shooting stars'. They are

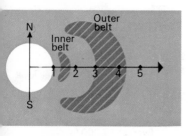

(*Above*) Auroral displays can have various striking forms and colours.

(*Left*) The inner and outer Van Allen radiation belts encircle the Earth in the plane of the equator. The horizontal scale is marked off in radii of the Earth. The location of the belts was charted by the satellite Explorer 4.

small particles of matter entering the atmosphere at speeds up to 45 miles per second and burning up from friction at heights of 50 to 100 miles. Occasionally larger masses known as *meteorites* reach the ground. Often they break into fragments in the air, with accompanying bangs.

## The Moon

The origin of the Moon is something of a mystery. One view was that the Moon was a part of the Earth which broke away, leaving a scar in the Earth's crust now filled by the Pacific Ocean. In modified form this idea still has some supporters. Another theory suggested that the Moon was a planet captured by the Earth's gravitational pull, and another that Earth and Moon happened to be formed together in space. The effect of the Earth's gravitational pull on the Moon has been sufficient to cause it to keep the same side towards us all the time; in other words the Moon rotates on its axis in the same time as its orbital period.

The Moon

The Moon appears to be very nearly a perfect sphere with no flattening at the poles, although it bulges up somewhat at the centre of the visible face. The average density of the Moon is less than the Earth, being about 3·4 times that of water, and it is unlikely that the Moon possesses a metallic core. The major features of the Moon's surface are easily seen in a small telescope or binoculars, and some features can be seen without optical aid. The features of the far side were unknown until space probes travelled around the Moon. Since the Moon moves faster at perigee than apogee, it is possible to see a little

(*Above*) Craters in the Alphonsus area. (*Opposite*) The Mare Imbrium bordered by the lunar Apennines (top left) and the Alps (bottom left). The large crater near the centre is Archimedes.

more than half the surface, at one time or another.

The main features of the Moon were first seen in 1609 by Galileo. Basically they fall into three categories: dark plains, ring-like craters, and mountain ranges. The early telescopic observers thought that the dark plains were seas and named them accordingly. Nowadays we know that the Moon has no surface water, but the names have been retained. The craters have been named after famous men, not all of them astronomers, though such familiar names as Ptolemy and Newton naturally appear. The craters range in diameter from giants like Bailly (180 miles) and Clavius (150 miles) down to the smallest resolvable size. Although the craters look deep when filled with shadow, their profiles are actually quite shallow and most craters more closely resemble saucers than cups. The main mountain ranges are spectacular: the lunar Apennines run for about 400 miles along the south side of the dark plain known as the Mare Imbrium (Sea of Showers) and attain heights of 18,000 feet. The Doerfel Mountains, near the lunar south pole, rise to 25,000 feet. Lone peaks are few, though excellent examples are Piton and Pico which stand in solitary splendour on the Mare Imbrium. The plains, usually approximately circular, range from the well defined Mare Crisium (200 miles diameter) to the Oceanus Procellarum which occupies a large part of the eastern section of the disc.

There are many lesser features worthy of note. Most immediately obvious, perhaps, are the bright rays which appear to emanate from some craters, such as the superb 56-mile crater Copernicus. Some of the rays from the crater Tycho travel virtually right round the Moon. Presumably these are the result of some sort of matter ejected from the craters, but their nature remains a mystery. The lunar surface is peppered with chains of little craterlets, and even large craters sometimes form short chains (Ptolemy, Alphonsus and Arzachel, near the centre of the disc, are an example). There are many winding valleys, or rills, as well as fault lines, scarps, ridges and other geological formations. An ex-

Hot magma beneath the crust

(*Opposite page*) Crater chains around the crater Copernicus. Bright rays appear to diverge from its central area.

(*Above*) The Triesnecker clefts

(*Above right*) Lunar domes in the Hortensius area

(*Right*) A possible crater-forming process

Crust forced up into dome

Crust ruptures—gas and material ejected

cellent example of a lunar fault is the straight wall some 60 miles long on the Mare Nubium. Of considerable interest are the low swellings, or lunar domes, which range in size up to tens of miles across.

After the 1830s, interest in the Moon lapsed, as it was thought that the Moon was a dead world where nothing ever happened. However, amateur observers continued to do a great deal of serious observing of the Moon and reports began to come in of alleged minor changes on the surface. The first really well-documented sign of surface activity on the Moon

Collapsed dome remelts in magma

Final appearance

American Ranger spacecraft over the Moon

was the observation of gaseous emission in the crater Alphonsus by the Russian observer Kozyrev in 1958. Since then a close watch has been kept on areas suspected of such outbreaks, and it is certainly true to say that there is minor localized activity of some sort on the Moon.

The theories of the origin of the craters have been many and varied. It was once suggested that the inhabitants of the Moon had had a devastating war, the craters being formed by bomb blasts. Nowadays, there are only two serious theories; either the craters were formed by some sort of internal volcanic activity or they were caused by the impact of meteorites in the past. Both types of crater are found on the Earth and both types must be found on the Moon. The question is which was the major crater-forming process.

What does the lunar surface look like? Basically there appear to be two principal types of rock, the dark rocks, often known as *lunabase* which form the Maria (seas) and the bright rocks, *lunarite,* forming the highlands, crater rims and so on. Earth-based observations gave some indications of the nature

Russian soft-landing Moon probe Luna 13

of these rocks. For example, by analyzing light reflected from them it was found that the lunabase areas were similar to lava flows on the Earth. In fact they look very like lava flows. It was thought that lunarite would be a lighter rock, perhaps of a honeycomb structure.

Whatever types of rock there may be on the Moon, they are subject to tremendous erosive forces. For one thing, temperatures range from over 100°C during the lunar day

Photograph taken by Ranger 9, 775 miles above Moon, showing parts of Ptolemaeus and Alphonsus.

down to around —130°C at lunar midnight, the extremes of temperature causing rocks to crack and splinter. The rapid temperature changes are largely due to the fact that the Moon has no atmosphere whatsoever (or at any rate, any atmosphere is less than 1/10,000 that of the Earth's) and so heat which reaches the lunar surface unimpeded during the lunar day, radiates away rapidly at night. The lack of atmosphere means that the surface is exposed to all sorts of radiations as well as countless meteorite impacts. In view of these erosive forces it was suggested that the Moon's surface might be covered with a thick layer of dust particles possibly hundreds of feet deep forming a sort of lunar quicksand into which any spacecraft would sink without trace.

The first successful Moon shots were the Lunik series of 1959. The Americans had many unsuccessful attempts, but really useful results were sent back by Rangers 7, 8 and 9 in 1965. These spacecraft carried television cameras and sent back thousands of close range photographs of the surface, before finally crashing into the Moon. These photographs were surpassed by those sent back by the Orbiter series of

Lunar surface seen from Lunar Module.

The Earth seen from lunar orbit.

probes which were sent into low level orbits around the Moon. In 1966 the Russians successfully soft-landed the craft Luna 9 on the surface. It did not sink and so immediately disposed of the deep dust theory. This was followed by the American Surveyor 1. The Apollo 11 and 12 flights of 1969 brought back superb colour photos taken by men on the Moon and samples of lunar rock.

What have the various Moon shots told us? The surface seems to consist, particularly in the highlands, very largely of rubble, ranging from huge blocks of rock downwards. There is a great deal of what would be called dust, but the surface layer has quite sufficient strength to bear the weight of a spacecraft since it is composed of a mixture of dust and all sizes of rubble mixed together. There is certainly no layer of lunar quicksand! However, neither the Orbiter nor the soft-landing craft nor the manned landings have been able to decide on the major crater-forming process.

Eventually a manned base will be established on the Moon. The Moon will make an excellent base for astronomical research—its lack of atmosphere will allow scientists to study all sorts of phenomena which could not be studied on Earth.

The planet Mars as seen telescopically

## MARS

Mars was the mythological god of war, and the blood-red colour of the planet is certainly symbolic of this. The first of the superior planets, Mars is one of the most interesting members of the solar system since in many ways it is quite similar to the Earth and may even support some form of life.

Mars moves round the Sun at an average distance of 141,500,000 miles, but because of the relatively large eccentricity of its orbit (0·09), its distance can vary from 129 million miles at perihelion to 155 million miles at aphelion. Thus the distance at opposition between the Earth and Mars can vary considerably: at best Mars can come as close as 35 million miles, but at a bad opposition it can be 60 million miles away. The apparent size of the planet at opposition ranges between 25 seconds and 14 seconds, while at superior conjunction it shrinks to 3·5 seconds. The planet is then up to 250 million miles distant. When Mars is at a good opposition, it can out-shine all the planets except Venus. Normally though, Jupiter is brighter than Mars.

Mars is a small world, only 4,200 miles in diameter with a mass of about one tenth that of the Earth, travelling around the Sun once every 687 days. The synodic period between successive oppositions is 780 days, so that Mars only comes into opposition every alternate year and thus opportunities to study the planet closely are not too common. However, study of the disc is rewarding as there are many well-defined features visible. The general colour of the planet is reddish, but there are many dark patches which the first telscopic observers took to be seas. Unfortunately there are definitely no open tracts of surface water on the planet. The reddish areas appear to be deserts of a sort. There are white caps at the poles, reminiscent of the Earth's ice caps.

Since Mars has an escape velocity of 3 miles per second it has only been able to retain a relatively thin atmosphere. Current estimates put the atmospheric pressure at only a fiftieth of the Earth's. The atmosphere can be analyzed spectroscopically from the Earth and various sorts of clouds can be seen.

One of the most prominent of the dark markings is a triangular patch known as Syrtis Major, which was first shown

(*Left*) Dates of oppositions of Mars. (*Right*) Size of Mars at various oppositions

(*Left*) Christiaan Huyghens made this drawing of Mars in November 1659.

(*Below*) The same area, the region of Syrtis Major as seen from Yerkes Observatory in September 1909

in a drawing made by the Dutch observer Huyghens in 1659. (This drawing was in fact used much later to help calculate a very precise value of the rotation period of Mars.) About the same time, Cassini first detected the polar caps. Herschel in the late eighteenth century made many drawings of the planet and deduced its rotation period as 24 hours 39 minutes 22 seconds. The value accepted today is 24 hours 37 minutes 23 seconds. About 1840, Beer and Madler produced a map of Mars, and since then Mars has been pretty intensively observed, although until recent years mostly by amateurs.

What sort of temperatures prevail on Mars? Since it lies further from the Sun than the Earth does, we would expect it to be colder, but as it possesses an atmosphere, albeit a rather thin one, it should not become as cold at night as the Moon. The planet's 24½-hour rotation period also implies that temperatures cannot drop too far or too suddenly. It now seems that the equatorial temperature of Mars can be as high as 25°C (70° Fahrenheit), although it drops to below −80°C at night. The climate is thus rather extreme, but not intolerable for forms of life which do not need a dense atmosphere. Although the Martian atmosphere is rarefied by our standards—roughly equivalent to our atmosphere at an altitude of about twenty miles—it is still sufficient to stop most meteors and minimize harmful radiations.

Virtually all astronomers are agreed that the reddish-ochre areas are deserts, but not composed of sand like the Sahara. The colour seems to be due to a deposit of coloured minerals, either felsite, an igneous material, or limonite, a form of iron oxide (the 'rusty desert' theory). The dark areas remain something of a mystery; some astronomers maintain that they are due to certain types of water-absorbing (hygroscopic) salts, and others, as we shall see later, believe that they are areas covered with a form of vegetation. There is no surface water, but it has been suggested that there might be some underground.

Clouds have often been observed in the Martian atmosphere, although they are not really analogous to our own rain clouds – rainfall must now be unknown on Mars. The Martian clouds fall into three main categories: high-level ('blue clouds'), intermediate level ('white') and low-level ('yellow'). The blue clouds are so called because they are best seen in short wave, that is, blue, light; in other words, they reflect more blue light than other wavelengths. They seem to lie at heights of the order of 50 miles, but this is by no means certain. Their nature is uncertain; ice

Map of Mars

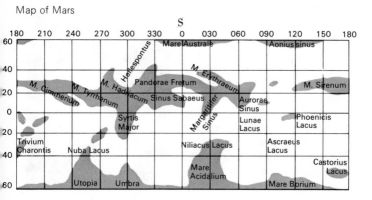

crystals were suggested at one time, but this now seems highly unlikely.

The white clouds are more readily seen and can occasionally become as prominent as the polar caps. They seem to lie at heights between 5 and 15 miles. There is more justification for suggesting that these clouds are fine ice crystals. Perhaps they are similar to some of our own cirrus clouds, which on Earth are usually forerunners of bad weather. Similar too is a whitish haze sometimes seen along the line of the Martian sunrise. It is tempting to compare this to terrestrial morning mist, but it is doubtful if there is enough water vapour on Mars to cause anything like a normal mist.

More prominent, and often obscuring large areas of the planet, are the yellow clouds. These are low level phenomena, almost certainly due to dust storms, material being blown up from the 'deserts'. The opposition of 1956 was the last really close one (the next will be 1971) and the surface features should have been well seen. However, in September and October that year virtually no detail could be seen because of planet-wide yellow obscurations.

Mars has two tiny moons, Phobos and Deimos ('fear' and 'terror', suitable companions for the god of war) both of which were discovered in 1877 by the American Asaph Hall. Both are only a matter of about 10 miles in diameter, Deimos orbiting the planet in 30 hours 18 minutes, while Phobos goes round in the astonishingly short period of 7 hours 39 minutes. Thus, to a Martian observer, Phobos would rise in the west.

The Martian polar caps are a source of great interest and show considerable variations throughout the planet's seasons. The seasons on Mars are similar in nature to those of the Earth. For one thing, the inclination of Mars' axis is 25·2 degrees as compared with the Earth's 23·5 degrees. The Martian southern summer occurs near perihelion too. The result is that Martian seasons differ from our own only in that they are more extreme and about twice as long. During the northern winter (southern summer), the north polar cap grows in extent and the south polar cap shrinks, sometimes vanishing altogether. In the northern summer, the situation is

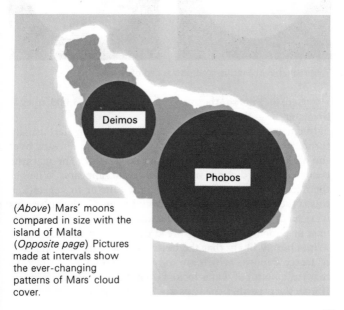

(*Above*) Mars' moons compared in size with the island of Malta (*Opposite page*) Pictures made at intervals show the ever-changing patterns of Mars' cloud cover.

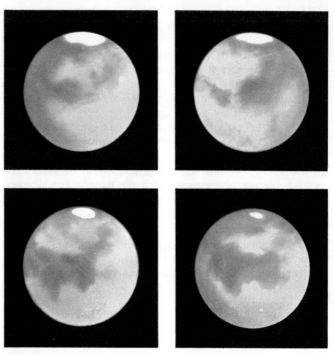

Seasonal changes in the south polar cap of Mars

reversed, and the southern cap often expands to 3,000 miles in diameter.

In spring, as a cap begins to decline, a dark 'collar' usually forms around it, and later a wave of darkening spreads over the dark areas, spreading towards the equator. The marking becomes more distinct and more colour appears. Whether the dark areas are due to hygroscopic salts or to vegetation, this is the sort of behaviour we would expect if the polar caps were made of ice which melted and evaporated to form water vapour which would then spread over the dark areas.

It is highly likely that the polar caps of Mars are composed of frosty deposits which are only a few millimetres thick, so that the amount of water locked up in the caps is pretty small. It may even be that the caps are due to solid carbon dioxide.

They are certainly not thick sheets of ice like the Earth's caps.

In 1965, Mariner 4 successfully reached the area of Mars, made measurements, and sent back a series of remarkable close-up photographs of the surface. These photographs clearly showed that the surface is covered in craters ranging up to about a hundred miles in diameter. The rims of these craters rise to heights of up to a hundred yards above the surface, while the crater floors lie many hundreds of yards lower. They seem to be highly eroded by the action of the Martian atmosphere. An interesting feature is that some of the crater rims appear to be covered with frosty deposits in areas experiencing winter. Mariner 4 has not answered what for most people is the most intriguing question: *Is there life on Mars?*

This photograph taken by Mariner 4 clearly shows craters on Mars.

Ever since astronomers began to study its surface in detail, Mars has been considered the one planet (apart from the Earth) in the solar system in which life, possibly similar to our own, might exist. It was suggested that Mars might be the home of intelligent beings, possibly far more advanced than ourselves, and such stories as H. G. Wells' *War of the Worlds* helped to arouse public interest in the matter. Nowadays it is still thought that life of a very basic sort may exist on the planet but super-civilizations and flying saucers get no support from serious astronomers.

A tremendous stimulus to interest in Mars was given with the discovery by Schiaparelli in 1877 of a pattern of linear dark markings on the surface. Schiaparelli called these features 'canali' or 'channels'. However, this has been translated as 'canals', which immediately suggests that they are of artificial construction. Percival Lowell founded the Flagstaff Observatory in the USA principally to study the Martian canals and devoted about twenty years from 1894 onwards to these observations. Lowell mapped hundreds of canals and became convinced that they were part of an enormous irrigation system constructed by the inhabitants to combat the water shortage on the planet. Sometimes canals would appear

Until recently astronomers produced pictures showing 'canals' on Mars. Later studies have ruled out this theory.

doubled, and he put this down to the opening of reserve canals. Nowadays Lowell's canals are no longer accepted. Observers with better equipment have been quite unable to see this network. Some of the so-called canals do exist, but under high magnification are seen to consist of patches of dark material roughly aligned. The eye has a tendency to join up such features at the limit of visibility. The discovery of craters on Mars suggests that the canals may in fact be crater chains formed along lines of structural weakness.

For life as we know it to exist on Mars, the planet would have to have an atmosphere whose density was not too greatly different from our own. Another essential is an adequate supply of free oxygen, the gas essential for human and animal life, and water. Mars has a very thin atmosphere, mostly nitrogen, with only about 0·1 per cent oxygen and very little water vapour. This, combined with the temperature range, rules out any of the advanced plant or animal life we know. The upper atmosphere cuts off harmful short-wave radiations (it is much less transparent to blue than red), but sometimes this layer can clear, exposing the surface to harmful rays.

Could any form of life exist on Mars? The answer is quite

Mars seen in blue light (*left*) and red light (*right*). As can be seen Mars' upper atmosphere is less transparent to blue.

possibly yes. Experiments have been carried out in laboratories in which various plants and organisms were subjected to simulated Martian conditions. Most plants died almost immediately, but some of the very primitive and basic ones survived a little longer. Certain bacteria lived on, not very happily admittedly, for considerable periods. This is more significant than it seems because it must be remembered that these bacteria survived without having had time to adapt from Earth conditions. If some form of lowly life once had a foothold on Mars, there is every chance that it would be able to develop to a small degree simply because it would be used to the environment. The behaviour of the dark markings on Mars is the sort of behaviour that one would expect of plant life, so there is circumstantial evidence at least of life on the planet. On the Earth, such hardy little plants as lichens which live quite happily on rocky mountain peaks certainly endure pretty unfavourable conditions. Any Martian plant life is unlikely to be more advanced than the lowly lichen. One thing is certain: animal and intelligent life in any way remotely like our own can be ruled out. The issue of plant life on Mars may be solved by unmanned space probes, but it is quite possible that the answer will have to await manned explora-

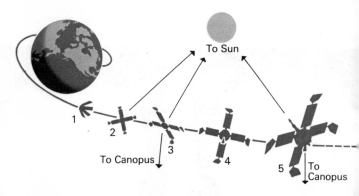

The path of Mariner 4 towards Mars. 1 Mariner unfolds solar panels. 2 Solar panels are turned toward Sun. 3 Sensor is focused on guide star Canopus. 4 Rocket fired to adjust path at mid-course. 5 Mariner is realigned with Sun and Canopus.

tion of the planet.

Although regarded as pure fiction ten years ago, a manned flight to Mars is certainly a possibility within the next decade or so. No one can be quite sure what the rate of progress will be. One serious problem which must be overcome is the question of decontaminating spacecraft and spacemen so that they do not take any earthly organisms to Mars. Likewise there is a danger of a returning craft accidentally introducing unknown Martian organisms, if they exist, to the Earth. At any rate, Mars appears to be a more favourable place on which to set up a base than the Moon. The temperature is better and the atmosphere offers some protection from meteors and harmful radiations. It is certainly the most welcoming planet in the solar system, apart from the Earth.

Artist's impression of the surface of Mars

## THE MINOR PLANETS

The planets of the solar system fall into two distinct groups: the four inner planets, Mercury, Venus, the Earth and Mars, and the five remaining outer planets, Jupiter, Saturn, Uranus, Neptune and Pluto. There is a distinct gap between Mars and Jupiter. Kepler went so far as to say, 'Between Mars and Jupiter, I place a planet.' In 1772, Bode of the Berlin Observatory drew attention to a curious relationship between the distances of the planets. This relationship, now known as Bode's law, is as follows.

Take the numbers 0, 3, 6, 12, 24, 48, 96, 192, where, apart from the first two, each number is double the previous one. Adding 4 to each we get 4, 7, 10, 16, 28, 52, 100, 196. As the table shows, if we take the Earth's distance from the Sun as being 10, and work out the relative distances of the other planets, Bode's law predicts the distances surprisingly accurately.

The law does not hold too well for Neptune and Pluto, and as we shall see later, Pluto's orbit at least is distinctly curious.

A glance at the table shows

Uranus 19

Saturn 10●

Jupiter 52

? 28

Mars 16

Earth 10

Venus 7

Sun

Mercury 4

| Planet  | Distance by Bode's law | Actual distance |
|---------|------------------------|-----------------|
| Mercury | 4                      | 3·9             |
| Venus   | 7                      | 7·2             |
| Earth   | 10                     | 10·0            |
| Mars    | 16                     | 15·2            |
| ?       | 28                     | —               |
| Jupiter | 52                     | 52·0            |
| Saturn  | 100                    | 95·4            |
| Uranus  | 196                    | 191·8           |

that there is no planet corresponding to the distance 28 predicted by Bode's law, that is, in the gap between Mars and Jupiter. If such a planet existed, it would have to be pretty small and faint to have escaped detection for so long.

In 1800 six European astronomers began a systematic hunt for the missing planet. The group became known as the 'celestial police', with the Hungarian Von Zach as secretary.

(*Opposite*) Bode's law indicates that there should be a planet between Mars and Jupiter.

(*Below*) A minor planet can be detected by a sequence of photographs (*a, b, c*) or, if it is close, by the streak it makes on a long exposure (*d*).

c

They were, however, forestalled by the discovery of a moving body on January 1 by Piazzi in Sicily. Piazzi wrote to Von Zach telling him of his observations and although by this time the possible new planet was too close to the Sun to be seen, the German, Gauss was able to calculate an orbit and predict the planet's reappearance. The following year the planet duly reappeared and was confirmed. Ceres, as the new planet was called, was a very small world, less than 500 miles in diameter.

The small size of Ceres led the celestial police to think that

(*This page*) Some minor planets compared in size with the British Isles

(*Opposite page*) Smaller asteroids are just lumps of rock in orbit about the Sun.

Flora

Vesta

Ceres

there might be other small planets in roughly the same region, and so they continued their search. Their patience was rewarded in 1802 when one of their number, Olbers, discovered a second minor planet, smaller and more distant than Ceres, which was named Pallas. Olbers suggested that the two planets might be fragments of a larger planet which had broken up for some reason. A third minor planet, Juno, was found in 1804, and a fourth, Vesta, in 1807. Vesta was the brightest of the four and can occasionally be seen with the naked eye as a very faint speck.

After this the celestial police gave up their search. However, in 1845 the German amateur Hencke discovered a fifth, Astraea, and a sixth two years later. By 1850 twelve minor planets were known and at the present time over 2,000 have had orbits calculated. Of the brighter minor planets, the following can be readily seen with a small telescope: Ceres, Pallas, Juno, Vesta, and amongst others, number 7 Iris and number 9 Metis. The first four are also the largest, their diameters being as follows: Ceres, 480 miles; Pallas, 300 miles;

Juno, 120 miles; and Vesta 240 miles. No other minor planets are known with diameters in excess of 100 miles and most are a matter of a few miles across.

It seems quite likely that there is no difference between a very large meteorite and a very small minor planet; in fact there may be a continuous distribution of sizes from minor planets down to meteorites. They are both solar system debris.

How were the minor planets formed? Although the larger ones look spherical, the smaller ones are irregular in shape,

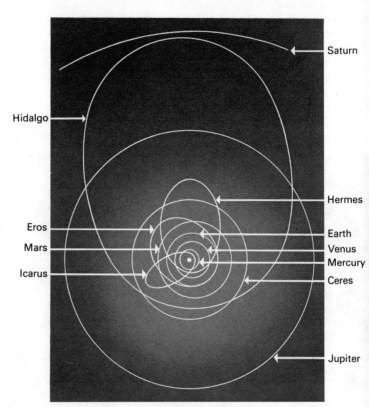

(*Above*) Orbits of some minor planets

(*Below*) The main asteroid zone lies between Mars and Jupiter.

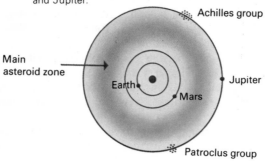

just like enormous rocky fragments. (It is hard to see the shape of a minor planet, but an irregular body rotating on its axis will vary in brightness as the minor planets tend to do.) Thus they look superficially like fragments of a larger planet which once orbited between Mars and Jupiter. Why a planet should explode in this fashion is not clear. The total mass of the minor planets is still very low, and so the original planet must have been quite small itself. Another theory is that they are simply material which did not form into planets.

Consideration of the orbits of the first three minor planets led Olbers to suggest that they were fragments of a planet which exploded. Celestial mechanics showed that if a body orbiting round the Sun exploded in this fashion, then the orbits of the fragments would more or less intersect at one point even though they were greatly different otherwise, this point being the place where the disruption took place. Now the orbits of Ceres, Pallas and Juno did seem to intersect at one point, in the constellation Virgo in fact, and so Olbers concentrated his search for further minor planets in this region. His hypothesis gained support from the fact that he discovered Vesta there. However, calculations soon showed that Vesta's orbit did not intersect the others. Now that we know that there are thousands of asteroids with non-intersecting orbits, one might think that Olbers' hypothesis was wrong. However, the minor planets' orbits are subject to perturbations by the planets, particularly Jupiter, and so it would be extremely surprising to find intersecting orbits after the (presumably) thousands of millions of years which have elapsed since the minor planets were formed. The question remains open.

The orbits of minor planets vary considerably. Ceres moves in a nearly circular orbit about 270 million miles from the Sun, while at the opposite extreme the minor planet Hidalgo has an orbit so eccentric that at aphelion it passes just outside the orbit of Mars at perihelion, but is almost as far out as Saturn at aphelion. Whereas most ordinary planets move in planes pretty close to that of the Earth, Pallas, for example, has an orbit inclined at 34 degrees to the Earth's. Icarus suffers tremendous extremes because, although at aphelion it recedes to a distance of 180 million miles from the Sun, at perihelion it

(*Left*) The orbit of minor planet Hermes almost intersects that of the Earth. (*Right*) Size of Eros compared with Malta

passes within 20 million miles of the Sun, much closer than Mercury does. The surface of Icarus must then be red hot.

Jupiter has collected minor planet 'families' of its own which move in the same orbit as itself. There is no danger of collision, though, because one family, the Greeks, lies 60 degrees ahead of Jupiter, while the other family, the Trojans, lies 60 degrees behind because of the combined pull of the Sun and Jupiter.

Since some of the minor planets pass closer to the Sun than the Earth does, it follows that their orbits must cross the Earth's. Normally they pass millions of miles above or below the Earth's orbit, but some come much closer. In 1899, minor planet 433 was discovered by Witt at Berlin, and named Eros. Eros is an elongated specimen about 15 miles by 4 miles, which approached within 17 million miles of the Earth in 1931. Careful measurements of its motions enabled a more accurate value of the astronomical unit (the distance from the Earth to the Sun) to be calculated. It is possible for it to approach within 15 million miles and its next close approach is in 1975. In 1932, two more minor planets were discovered which approach much closer, Amor and Apollo. Termed 'Earth-grazers' for obvious reasons, they came within 10 million and 7 million miles respectively. The record soon fell to Adonis which passed us at a mere 1,300,000 miles.

The year 1937 saw the discovery of an even more interesting Earth-grazer named Hermes, only about a mile in diameter. It came as close as 485,000 miles, and if its orbit is not greatly perturbed during its travels, it may pass between the Earth and the Moon. The close approach in January 1938 was seized upon by the press as a world disaster narrowly missed, but there was no danger of a collision. For such a collision to occur two things are necessary: (*a*) the orbital planes of the minor planet and the Earth intersect, and (*b*) both bodies must be at exactly the same point at the same time. No minor planet is known satisfying condition *a*, and the chances are millions to one against this being so. Even if *a* were true, *b* is

Seen from a minor planet the Sun would be intense.

such a remote chance that it is not worth worrying about. Nevertheless, if such a collision did occur with a sizable minor planet, the result would be devastating.

Various suggestions have been made as to possible uses for minor planets. They could be used as manned or unmanned observatories—Icarus would be a case in point, unmanned of course—or they might even be mined for any minerals they contain. Anyway, they are interesting objects.

# THE GIANT PLANETS

Jupiter and Saturn are the giants of the solar system, and Jupiter in fact is much more massive than all the other planets put together. Jupiter was the king of the gods that the ancient Greeks believed to live on top of Mount Olympus and such a title surely befits the largest of the planets. Saturn was Jupiter's father and the god of time and fate, his great age being somehow suggested to astrologers by the planet's yellowish colour. Astrologers attributed malign powers to Saturn and predicted dire consequences for anyone born under the influence of this planet.

## Jupiter

Jupiter is a truly enormous world, eleven times greater in diameter than the Earth with an equatorial diameter of 88,700 miles. The volume of the planet is quite sufficient to hold 1,300 Earths. However, Jupiter's average density is remarkably low, about 1·3 times greater than water, so that its mass is equivalent to a mere 317 Earths.

Jupiter revolves about the Sun in a period of 11 years 10 months 3 days at an average distance of 483 million miles. Its orbit has an eccentricity of 0·048, considerably more than the eccentricity of the Earth's path, and so at perihelion Jupiter is 460 million miles from the Sun as against 507 million at aphelion. Jupiter's distance from the Earth can vary from 390 million miles at a good opposition to 600 million at superior conjunction, the angular diameter ranging between 30 and 50 seconds of arc. This variation in size is not nearly as great as for any of the planets we have considered so far, and Jupiter's brightness does not vary greatly either. Usually Jupiter is the brightest of the planets, except when Venus is well placed and occasionally when Mars is at a particularly favourable opposition. The planet's synodic period is 399 days and so successive oppositions occur roughly a month later each year. Jupiter is, in fact, a very easy planet to observe since it presents a large disc and is visible all the year round apart from a month or two near superior conjunction. Even then its apparent size is greater than Mars at its best. Small telescopes will show a lot of detail.

(*Above*) Jupiter and Saturn compared

(*Below*) Jupiter and Earth compared

The cloud belts of Jupiter

The first person to observe Jupiter telescopically was Galileo in 1609, and even with his tiny telescope quite a lot could be seen on the planet. One immediately obvious fact is that the disc is far from circular; Jupiter is flattened at the poles and bulges at the equator, so that whereas its equatorial diameter is 88,700 miles, the diameter from pole to pole is only 83,800 miles, a difference of nearly 5,000 miles. The Earth too is flattened at the poles, but only by a matter of about 20 miles. This flattening (the shape is called an oblate spheroid) suggests that the planet is not truly solid in the sense that the inner planets are and it seems that the surface we see is the top of a sea of clouds. Jupiter has a very fast rotation rate: it rotates on its axis in less than 10 hours, and when we consider the circumference of the planet we find that the equatorial regions are spinning round at something like 28,000 miles per hour (more than the Earth's escape velocity). The equatorial bulge is caused by the centrifugal force generated by this rapid rotation.

Through the telescope Jupiter is seen to be covered with a

system of cloud belts parallel to its equator. These belts are constantly changing in width and visibility and tremendous activity and changes take place within them. The rotation of the planet can be observed by watching the motion of projections and humps on the belts, and a few minutes' observation with a large telescope will show this motion quite clearly. Careful study of the belts has shown that the equatorial regions of the planet rotate slightly faster than the areas nearer the poles, and various currents have rotation periods of their own. The equatorial regions are taken as rotating in 9 hours 50 minutes (and are termed System 1) while the rest of the planet is assumed to rotate in 9 hours 55 minutes (System 2). (See illustration on page 152.)

The temperature of Jupiter is very low, of the order of $-140°C$, and the surface gravity (the force acting on a body on the surface) is $2\frac{1}{2}$ times greater than the surface gravity of the Earth.

Because Jupiter is so massive, its escape velocity is 37 miles per second, and so it has been able to retain all the gases, including hydrogen, the fastest moving of them all. It seems that hydrogen is the principal constituent of the planet. Spectral analysis of the cloud belts has shown these to be composed principally of ammonia and methane, both of which are compounds of hydrogen. Ammonia (chemical formula

Observing the motion of irregularities in the cloud belts of Jupiter shows the planet's rotation.

NH$_3$) is composed of hydrogen and nitrogen combined, while methane (CH$_4$) is made up of hydrogen and carbon. Methane when mixed with oxygen is explosive, and is responsible for many terrestrial mine disasters.

What is the complete structure of Jupiter? There are two main theories at the moment, the first being the work of Wildt in the USA. Wildt proposed that the planet has a rocky metallic core about 38,000 miles in diameter around which is wrapped a layer of ice 17,000 miles thick. On top of this lies the 'atmosphere', which is composed principally of hydrogen. It is interesting to note that the gases in the lower regions of this

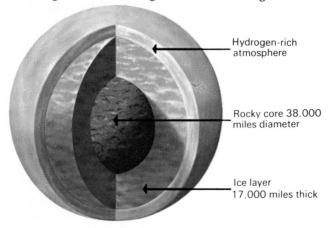

Hydrogen-rich atmosphere

Rocky core 38,000 miles diameter

Ice layer 17,000 miles thick

Wildt's model of Jupiter

atmosphere would be subject to enormous pressures and would therefore behave much more like a solid than our idea of a gas.

Ramsey, in England, has proposed an alternative view which has received much support. Basically he suggests that Jupiter is composed almost entirely of hydrogen. The atmosphere consists, as in Wildt's theory, of a mixture of ammonia, methane and hydrogen. Pressure increases rapidly with depth and is soon sufficient for hydrogen to behave as a solid. By the time a depth of 5,000 miles is reached, the pressure has shot up to something like 800,000 atmospheres and the hydrogen behaves like a metal. At the centre of the planet the metallic hydrogen would have a density about 3·7 times the density of water.

Thus in Ramsey's model, Jupiter consists of a core of metallic hydrogen 76,000 miles in diameter surrounded by a layer of solid hydrogen some 5,000 miles thick. On top of this core lies a relatively thin atmosphere culminating in the visible surface. There could also be a layer of liquid hydrogen on top of the solid region, forming an immense hydrogen ocean.

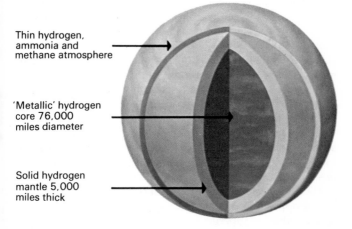

Thin hydrogen, ammonia and methane atmosphere

'Metallic' hydrogen core 76,000 miles diameter

Solid hydrogen mantle 5,000 miles thick

Ramsey's model of Saturn

Many spots become visible on the cloud belts, but few persist very long. In a different category altogether is the *Great Red Spot,* which first became prominent in 1878, although it seems likely that Robert Hooke recorded indications of it as early as 1664. Lying in the south temperate belt, it developed into a brick-red oval about 30,000 miles long and 7,000 miles broad. After 1882, the spot began to fade, but the 'hollow' in which it seemed to lie remained faintly in view. Since that time the Great Red Spot has revived periodically. It vanished from view in 1959, but, since 1960, it has been quite prominent again.

Obviously something like this is no ordinary cloud, but astronomers have no clear idea what it can be. It may well be a solid, or semi-solid, lump of matter floating in Jupiter's atmosphere, possibly of solid ammonia with sodium mixed in, but it remains a mystery. Various currents, often characterized by small spots, exist in the Jovian atmosphere. In the past, a darkish zone known as the *South Tropical Disturbance* has sometimes been visible, which interacts with the Great Red Spot, overtaking it and dragging it thousands of miles out of position before the spot slips back again.

As befits the largest planet, Jupiter has a family of twelve moons, four of which were seen by Galileo, and which some observers with exceptional eyesight can see without telescopic aid. These four are Callisto (diameter 3,220 miles), Ganymede (3,200 miles), Io (2,310 miles) and Europa (1,950 miles). Thus two of them are bigger than the planet Mercury, though not as massive. It is fascinating to watch the motions of these satellites as they and/or their shadows cross Jupiter's disc, or as they are eclipsed by the planet. Mutual phenomena between the satellites also occur. The four bright satellites seem to be much more efficient at reflecting light than is our Moon. Jupiter's other moons are small, faint and difficult to observe.

The Danish astronomer Rømer observed that when Jupiter was at opposition satellite phenomena were earlier than predicted, while at superior conjunction they were late. He deduced that this was due to the time taken by light to travel the extra distance to the Earth, and so worked out a value for the velocity of light quite close to the presently accepted value.

In recent years it has become apparent that Jupiter is a source of radio waves, though the origin of these waves remains a mystery. Attempts are being made to correlate the radio signals with surface features, using mainly amateur observations, but these efforts have not met with much success.

Radio astronomers have also discovered that Jupiter has belts of charged particles in its magnetic field, like the Earth's Van Allen belts. Jupiter's magnetic field is probably a great deal stronger than that of the Earth.

This sequence of pictures made at intervals shows the development of a disturbance in the south equatorial belt of Jupiter.

Jupiter and some of its moons

## Saturn

Saturn is one of the most beautiful telescopic sights in the entire heavens, principally because of the system of rings, unique in the solar system, which surround the body of the planet. Saturn was the outermost planet known in pre-telescopic days and does not look particularly bright or interesting to the naked eye.

Saturn moves around the Sun at an average distance of 886 million miles in an orbit of eccentricity of 0·056, so that its perihelion distance is 840 million miles and its aphelion distance 938 million miles. Saturn next reaches perihelion in 1974. The planet's orbital period round the Sun is 29 years 167 days, while its synodic period is 378 days, so that successive oppositions are less than a fortnight later each year. Saturn's rotation period on its axis is rapid, about $10\frac{1}{4}$ hours with variations from equator to poles, and so it too is markedly flattened at the poles, even more so than Jupiter. Saturn's polar diameter is 67,200 miles as against its equatorial diameter of 75,100 miles.

The planet Saturn

(*Left*) From Saturn the Sun would appear as a very bright star

(*Below*) Hay's white spot on Saturn

The volume of the planet is sufficient to contain 740 Earths, but its mass is only 95 times that of the Earth and so its density is very low, 0·7 in fact, less than that of water. The surface gravity (assuming one could stand on the surface) is only slightly greater than the Earth's as this quantity depends on the mass of a planet but decreases with increasing planetary diameter.

The general appearance of the planet's disc is similar to Jupiter, but the cloud belts are fewer and less prominent. The most obvious belts are the north and south equatorial belts, the latter often appearing double, but little detail can be seen in the belts even with quite large telescopes. Spots are rare and there is nothing to compare with the Great Red Spot. The last really prominent spot was a white one in the equatorial zone discovered in 1933 by the amateur W. T. Hay. It was conspicuous enough to be seen in a 3-inch refractor, but soon dissipated and vanished. Another spot began to develop in 1962 but did not really become prominent.

The temperature on Saturn is around —180°C. Because its temperature is somewhat lower than that of Jupiter, more of the ammonia has been frozen out of Saturn's atmosphere than is the case with Jupiter, and thus spectroscopic results show a greater proportion of methane in the planet's atmosphere. It seems likely that Saturn is similar in general structure to Jupiter, being composed mainly of hydrogen. In Wildt's model the rocky metallic core is 28,000 miles in diameter, covered with an 8,000-mile ice shell, over which lies an atmosphere 16,000 miles thick. Ramsey's scheme gives Saturn a metallic hydrogen core of 25,000 miles diameter, overlaid by 8,000 miles of solid hydrogen and the atmosphere.

Early astronomers' notions of Saturn before Huyghens established in 1659 that Saturn had a ring system.

The rings are of course the most striking feature and were a source of great puzzlement to the early observers, including Galileo. His telescope was not powerful enough to show their real nature and he thought that Saturn had two attendant bodies, one on each side. As the appearance of the rings varies in cycles, Galileo was greatly confused to find that the 'attendants' had apparently vanished some years after his original observations. The existence of the ring system was finally recognized by Huyghens in 1659 with the aid of a superior telescope.

The ring system consists of three concentric rings nowhere touching the planet. The outermost ring, Ring A, is some 10,000 miles wide; Ring B, the brightest of the three, is 16,000 miles wide; and Ring C, known as the Crepe Ring, the dullest, is 10,000 miles wide. The Crepe Ring is so tenuous that the globe of the planet can sometimes be seen through it. Between Ring C and the body of the planet is a clear space some 9,000 miles wide into which the Earth could fit with room to spare. Between Ring A and Ring C is a gap 1,700 miles wide which

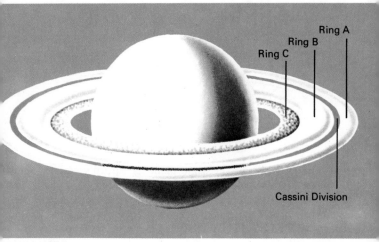

The rings of Saturn

was discovered in 1675 by Cassini and is known as Cassini's Division. Under good conditions, Cassini's Division can be seen with a 3-inch refractor.

The total diameter of the ring system is about 170,000 miles, but the thickness is probably no more than about 10 miles. The rings are certainly not solid. If they were, they would have been broken up by the planet's gravitational pull. They seem to be made up of myriads of small particles which are either tiny lumps of ice, or solid particles that are coated with ice.

Varying aspects of Saturn's rings. The lower pictures show how they appear from Earth at the two points in Saturn's orbit.

The rings lie in the same plane as Saturn's equator and the planet's axial inclination is 26·7 degrees. Thus it is easy to see that at one point in its orbit the south face of the rings will be on display, while half an orbit later, the north face will be on display. Halfway between these points, the rings will be edge-on to us. The rings thus vary in appearance from being quite wide open from our point of view, to being edge-on when they seem to vanish in all but large telescopes on account of their narrowness. The rings were last edge-on in 1966, and at the present time the south face is showing. They will be widest open again in 1973–4 and edge-on once more in 1979–80. Naturally, the brightness of the planet depends on the angle of the rings.

Saturn has quite a tally of moons, ten in all, the most recently discovered being Janus which was found by Dollfus in 1966. Janus travels round Saturn in a period of only 18 hours at a distance only 14,000 miles outside Ring A. Its discovery was facilitated by the fact that it lay so close to the rings which were then edge-on and were being intensively studied. Of

Saturn's other moons, the largest is Titan with a diameter of 3,000 miles. It is easy to see with a small telescope and was discovered by Huyghens in 1655. The other moons are fainter but several can be seen in moderate telescopes. Iapetus, another of the moons, is remarkable for its varying brightness. The divisions in the rings are due to the perturbing effects of the satellites on the ring particles.

Is there life on the outer planets? There is no chance of any life based on our pattern, but it is not totally inconceivable that some other form of life might exist (page 130). The answer will have to await exploration of the outer planets by spacecraft. At the present time we do have the ability to send unmanned space probes to Jupiter and Saturn but no manned flights are likely for some decades. Such a flight would have to last for many years.

A rocket as long as the largest ocean liner would be needed to send a manned spacecraft to Saturn with existing fuels.

## THE FRINGE OF THE SOLAR SYSTEM

Beyond Saturn lies the twilight zone of the solar system, where the Sun appears much smaller and paler than it does to us and where there lie the last three known members of the solar system, all of which were discovered telescopically. These are, in order of distance from the Sun, Uranus (father of Saturn), Neptune (the sea god) and Pluto (god of the underworld).

### Uranus

Throughout recorded history, five planets had been known, and no-one had seriously thought that there might be any more beyond the orbit of Saturn. Even after the advent of the telescope, this view persisted. However in 1781, William Herschel, then an amateur astronomer and telescope maker who made his living as a musician, was making systematic

The planet Uranus

| Earth 23° 45′ | Mars 25° 2′ | Jupiter 3° 1′ |

| Saturn 26° 75′ | Uranus 98° | Neptune 29° |

The axial inclinations of the planets

studies of the stars when he observed, as he said, 'one star which appeared visibly larger than the rest', which he took to be a comet. It soon became clear, however, that this was not a comet, but was a new planet moving well beyond the orbit of Saturn. The discovery of the planet, named Uranus by Bode, took the scientific world completely by surprise and Herschel was greatly honoured, being made the King's Astronomer. The planet had in fact been observed before on several occasions but had been mistaken for an ordinary star.

Uranus is exceedingly remote from the Sun, its mean distance being 1,782 million miles and its orbital eccentricity 0·44, giving a perihelion distance of 1,704 million miles and an aphelion distance of 1,870 million miles. The planet was at perihelion in 1967 and so is now moving out from the Sun towards aphelion which it will reach in the year 2011. Uranus has an orbital period of no less than 84 years.

A most peculiar thing about Uranus is the tilt of its axis of rotation. Most planets have their axis pretty well vertical to the plane of their orbits. Jupiter's axis is tilted a mere 3 degrees, while the Earth's is at an angle of $23\frac{1}{2}°$. The axis of

Uranus, however, is canted over at no less than 98 degrees and so its rotation can be regarded as being retrograde (that is, east west). Sometimes, therefore, we see Uranus pole-on and at other times its equator is presented.

Although by no means as large as Jupiter, Uranus is still a huge planet with an equatorial diameter of 29,000 miles and a polar diameter about 1,500 miles less. This seems to imply that Uranus has a structure similar to that of Jupiter and Saturn, and indeed the rotation period is similar at about 10 hours 50 minutes. The temperature is so low ($-210°$C) that nearly all the atmospheric ammonia is frozen, so that methane shows up predominantly. Again hydrogen would seem to be the principal component. Wildt's model gives Uranus a solid core of about 14,000 miles diameter, with a 6,000-miles ice layer and 3,000-mile atmosphere.

Uranus has a family of five satellites, two of which, Titania and Oberon, were discovered by Herschel. The English amateur Lassell found two more, Ariel and Umbriel, in 1851, while the faint Miranda was discovered by Kuiper in the USA in 1948. The sizes of these moons range from about 200 to 1,800 miles, and their distances from their parent planet from 76,000 to 364,000 miles. Since they orbit in the plane of the planet's equator, their orbits appear near circular when the planet is pole-on, and almost straight up-and-down lines when the equator is presented.

## Neptune

Soon after its discovery, it became apparent that Uranus was not behaving itself. Using old observations made before it was recognized as a planet, it should have been possible to work out an accurate orbit, but the planet refused to stick to its predicted path. Up until 1822 it seemed to move too fast, while afterwards it lagged. It began to look as if some unknown body was perturbing Uranus, and in 1843 a young English mathematician, John Couch Adams, began to work on the problem. By 1845 he had worked out the mass and position of the proposed planetary body and sent his results to G. Airy, the Astronomer Royal of the time, who however took no action.

Meanwhile in France, Urbain Le Verrier had begun a

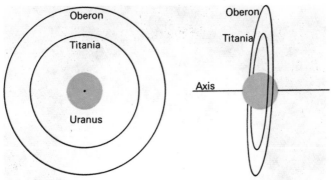

(*Above*) The moons of Uranus orbit in the plane of the planet's equator and thus their orbits appear circular when the planet is pole-on to the Earth.

(*Below*) Neptune was discovered by calculations based on the perturbation of Uranus' orbit.

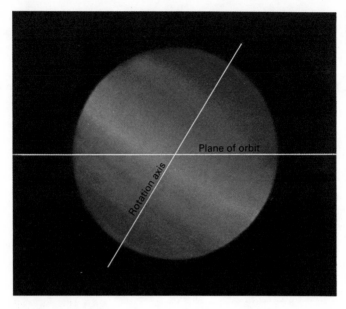

Rotation axis

Plane of orbit

The planet Neptune, showing cloud belts

similar calculation and produced very similar results to
Adam's values in 1846. Although Airy then instituted a
search, the planet was found by Galle and d'Arrest in Berlin.
However, the honour of the prediction must be shared by Le
Verrier and Adams.

The discovery of this planet, named Neptune, within a
degree of its predicted position was a major vindication of
Newton's laws of gravitation and dynamics as well as a great
personal triumph for the men involved. Neptune lies 2,739
million miles from the Sun on average and moves round in an
orbit of low eccentricity (0·007) in a period of 164¾ years.
From this distance, the Sun's apparent diameter is 30 times
smaller than it appears from the Earth and thus the Sun's disc
could only just be made out with the naked eye. The total
apparent brightness of the Sun would be only about a thousandth of its brightness as we see it.

Neptune is similar in size to Uranus, with an equatorial

diameter of 27,700 miles, but it is the more massive of the two, being about 17 times more massive than the Earth whereas Uranus is 15 times the Earth's mass. Neptune's density is 2·2 times that of water and so it is the densest of the four Jovian-type planets. Because of this it is much less flattened at the poles than the others, even although its rotation period is still fairly short, around 14 hours. This figure could be quite inaccurate as little can be seen on the planet's disc.

The temperature prevailing on Neptune are even lower than on Uranus and a value of around —230°C has been estimated. The Neptunian clouds seem to be composed largely of methane, this time with no detectable traces of ammonia. The planet's structure is certainly similar to the other Jovian planets, and Wildt has produced a model in which Neptune has a 12,000-mile rocky core, a 6,000-mile ice layer and a 2,000-mile gas layer. Ramsey maintains, as in the case of Uranus, that hydrogen is the principal constituent.

As far as we know, Neptune has only two satellites. The first of these is Triton, discovered a few weeks after the planet itself, and with a diameter of about 3,000 miles, one of the largest in the solar system. It moves in a near-circular retrograde orbit. The other moon, Nereid, has an extremely eccentric orbit.

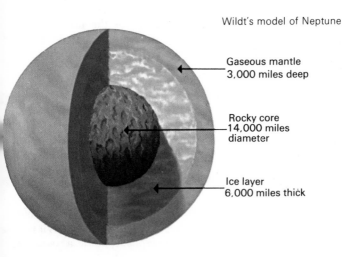

Wildt's model of Neptune

Gaseous mantle
3,000 miles deep

Rocky core
14,000 miles
diameter

Ice layer
6,000 miles thick

## Pluto

Pluto was discovered by comparing photographs taken 3 days apart.

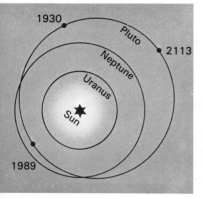

The orbits of the three outer planets

For about fifty years it was assumed that the discovery of Neptune had accounted for Uranus' erratic motions. However, other lesser discrepancies began to appear. Percival Lowell decided that there might be yet another planet causing these minor variations, even more remote than Neptune. Accordingly he began an Adams-Le Verrier type of calculation to find such a body. He predicted that 'Planet X' should have a mass about six times that of the Earth and move round the Sun at a distance of 4,000 million miles in a period of 282 years.

Lowell searched for many years, but it was not until 1930, fourteen years after his death, but the ninth planet was found by Clyde Tombaugh at Lowell's Observatory. Pluto, as it was called, was fainter than expected, but the calculations did not seem too far out. Pluto's average distance from the Sun is 3,666 million miles and its orbital period is 248 years.

However, it soon came to light that Pluto was far smaller than expected – current estimates give a diameter of about 3,000 miles – and unless it was improbably dense, it could not

Lowell at his telescope.

produce the required perturbations of Uranus. Another odd feature is the large eccentricity (0·248) of its orbit which causes Pluto's path to cross Neptune's so that for part of its orbit near perihelion it is not actually the outermost planet. Perihelion is next reached in 1989.

The probable low mass of Pluto has raised doubts as to whether it can be the body solely responsible for perturbing Uranus and has led some astronomers to suggest that there may be a tenth planet awaiting discovery. If so, it must be very faint, but it may be found some day.

Travel out as far as Neptune or Pluto is certainly beyond our capabilities at present with manned craft and is likely to remain so for a long time. A flight to Pluto and back could take many decades. From Pluto the Earth would be quite invisible, even telescopically.

## Planetary data

| Planet | Mean distance from Sun, in millions of miles | Sidereal period | Synodic period |
| --- | --- | --- | --- |
| Mercury | 36 | 88 days | 115 days |
| Venus | 67 | 224·7 days | 584 days |
| Earth | 93 | 365 days | — |
| Mars | 141·5 | 687 days | 780 days |
| Jupiter | 483 | $11\frac{3}{4}$ years | 399 days |
| Saturn | 886 | $29\frac{1}{2}$ years | 378 days |
| Uranus | 1,783 | 84 years | 370 days |
| Neptune | 2,793 | $164\frac{3}{4}$ years | $367\frac{1}{2}$ days |
| Pluto | 3,666 | $247\frac{3}{4}$ years | $366\frac{3}{4}$ days |

| Axial rotation | Diameter in miles (equatorial) | Brightest magnitude | Maximum surface temp. (in degrees Centigrade) |
|---|---|---|---|
| 59 days | 3,100 | −1·9 | 400° |
| 243 days | 7,700 | −4·4 | 350° |
| 23 h. 56 m. | 7,926 | — | 60° |
| 24 h. 37 m. | 4,200 | −2·8 | 25° |
| 9¾ hours | 88,700 | −2·5 | −140° |
| 10¼ hours | 75,100 | −0·4 | −160° |
| 10¾ hours | 29,600 | +5·6 | −210° |
| 15¾ hours | 27,700 | +7·7 | −230° |
| 6 d. 9 h. | 3,600? | +13 | ? |

## PLANETS OF OTHER SUNS

Are there any other planets in the universe, or is the solar system unique, and the Sun unique in having such a retinue? If we fully understood how the solar system was formed, it would be easier to answer such a question. What we do know is that the Sun is a very ordinary star. It is simply one of about a hundred thousand million stars in our galaxy, and a thousand million galaxies can be seen with large telescopes. Thus the universe which we can see contains something like a hundred million million million stars. Can the Sun be the only star in all these to possess a planetary system?

Jeans (page 14) proposed that the solar system was produced by a happy accident, when another star passed near to the Sun, and he thought that since the stars are so far apart, it was quite possible that nothing of this nature had happened anywhere else—within our own Galaxy at least. Jeans' theory has been abandoned, but Woolfson's is similar. However, we now know that stars tend to form in relatively compact groups and so there would be a definite chance of such interactions. Other theories imply that planetary systems tend to form naturally with any star and so planets accompanying stars might be the rule rather than the exception. Thus there may be thousands of millions of planets in the Galaxy; and even if Woolfson's theory is correct there should be tens or hundreds of thousands.

Even if this is the case, it would be quite impossible to observe such planets directly by means of telescopes as they would be far too small and faint. Remember, the nearest star is over 4 light years distant. However, there is one possibility for detection. Just as the Earth and Moon move

The wobbling motion of
61 Cygni indicates the presence
of a planet.

Possible size of 61 Cygni's planet compared with Jupiter

around the barycentre (page 66), so a star and its planetary companion(s) would move around their common centre of gravity. If the planet in question were sufficiently massive, motion around their centre of gravity would show up as a wobble in the star's motion. In 1944 it was discovered that the fainter component of the star 61 Cygni does just this.

61 Cygni, so called because it lies in the constellation Cygnus and was number 61 in Flamsteed's star catalogue, is visible to the naked eye and in 1838 became the first star to have its distance accurately measured. It lies about 11 light years distant and is a double star in which the components, both fainter than the Sun, lie quite far apart. The fainter of the two components, 61 Cygni B, shows a wobbling motion, and from measurements of the extent of the wobble, K. A. Strand deduced that the perturbing body responsible must have a mass about fifteen times greater than Jupiter. This may seem extremely massive, but it is far too small to be a star of any sort. The smallest known star has a mass more than 150 times that of Jupiter. It thus seems pretty certain that what is perturbing 61 Cygni B, is a dark body with no light of its own, in other words a planet.

Another such case is the star 70 Ophiuchi, also a double. It is not certain around which component of the pair the planetary body revolves. If it revolves about the major component, then its mass must be at least twelve times that of Jupiter, whereas if it revolves around the lesser component, its mass need only be equal to eight Jupiters.

In many ways the most interesting case is Barnard's Star, named after the astronomer who noticed its rapid motion. Barnard's Star is only 6 light years distant and quite small by stellar standards. Its companion, discovered by van de Kamp in 1963, was found to have a mass 1·5 times that of Jupiter. This body must certainly be a planet and in fact revolves around Barnard's Star in a period of 24 years, twice Jupiter's period around the Sun. Although we cannot detect

View of red giant star from lifeless planet

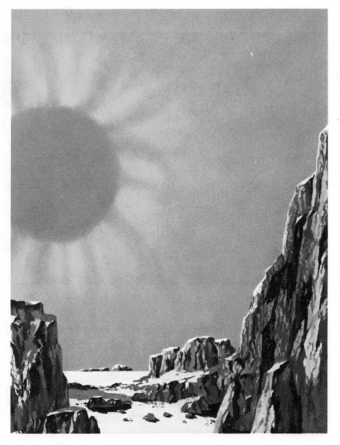

planets any smaller than this, it does not mean they do not exist. Quite probably, each of these giant planets is a member of a system containing smaller ones as well. Many of these planets must be strange and hostile to our form of life, but it is possible that somewhere there are planets more suited to life as we know it.

## Is there another Earth?

What interests many people is the possibility of finding an Earth-like planet, and many science-fiction stories have been woven around the possibility of there existing a planet

Group of galaxies

somewhere in the universe which is an exact replica of the Earth. There are too many variable quantities for this to be a possibility worth considering. What is possible, if planetary systems are as common as they seem to be, is the existence of planets where the conditions are similar to conditions on Earth, and to which our form of life could rapidly adapt. If life had gained a foothold on such a planet, it is possible that life closely parallelling our own could have developed.

What sort of conditions are necessary for life as we know it to develop? First of all of course a suitable planetary body is essential. Given this, then two vital conditions must be satisfied. The temperature must be neither too hot nor too cold, since intense heat breaks down organic molecules and severe cold prevents activity from going on. Too much short-wave radiation also upsets living organisms. The other pre-requisite is a suitable atmosphere sufficiently dense to give protection from radiation and meteorites and containing oxygen and water vapour in reasonable quantities.

Within our own solar system there is a 'region of tolerance'

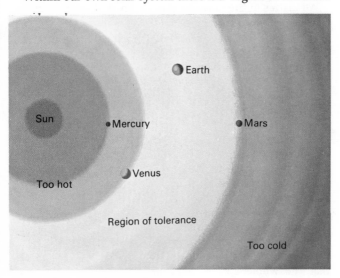

Region of tolerance in our solar system

where the amount of heat landing on a unit area of a planetary surface would be tolerable even to human life. This area extends roughly from the orbit of Venus to the orbit of Mars, and if the Earth orbited anywhere in this region, life could continue. However, Mars and Venus fail the second criterion. Venus' atmosphere is totally hostile, too hot, and with too high an atmospheric pressure for any earth organism to survive. The Martian atmosphere is too thin with too little oxygen and water vapour. However, Mars does not fail the test by much, and as we have seen, it is quite likely that life in some form has developed there. There are possibly two life-bearing planets in the solar system; the chances of there being others in the Galaxy are surely quite good.

## Other forms of life ?

So far we have been considering life quite similar to that on Earth, but it seems likely that life could develop along different lines. In fact, many people consider that life will develop wherever conditions are at all suitable for it.

All matter in the universe is made up of 92 basic elements. Our form of life is based on the element carbon, which has the property of forming the very complex atom-groups, or molecules, necessary for organic matter. The carbon atom is itself made up of a combination of three types of smaller particle, the proton, the neutron and the electron, and this structure is what gives it its properties. In living matter, many other elements are combined, but the basic structure depends nonetheless on carbon.

As far as we know, carbon is the most efficient element in its ability to form complex molecules, and it is fairly certain that any life-form we may encounter on other planets will be carbon-based, and thus basically similar in structure if not in shape to earthly organisms. However, the element silicon can also form fairly complex structures, and we cannot rule out the possibility that under conditions prevailing on some unknown planet a form of life might have arisen with this base. Another possibility, though it must be emphasized that the whole idea is controlled speculation rather than scientific prediction, is that a carbon-based life-form might be able to exist at very low temperatures in an environment in which water is replaced by liquid ammonia. In view of the conditions prevailing on Jupiter and Saturn, this is an interesting idea.

All we can say is that there certainly are other planets in the universe and it is quite probable that there are large numbers of these. It is therefore pretty certain that some form of life has developed on some of them, and intelligent life is a possibility we would be foolish to ignore.

| | Number of | |
|---|---|---|
| 1 | electrons | 6 |
| 1 | protons | 6 |
| 0 | neutrons | 6 |
| 1 | atomic number | 6 |
| 1 | atomic mass | 12 |

The hydrogen (*left*) and carbon atoms

(*Above*) The DNA molecule is thought to consist of two chains of alternating sugar and phosphate molecules in a double helix. This model shows a chain splitting and reforming to duplicate the DNA molecule. (*Right*) A simple bacterium. Such organisms may exist on other planets.

(*Above*) The blast-off of an Apollo mission to the Moon

(*Right*) Section through Apollo Command Module. The centre couch has been removed.

# RECENT DEVELOPMENTS AND FUTURE TRENDS

In September 1968, the Russian unmanned craft Zond 5 successfully flew round the Moon, returning safely to the Earth, splashing down in the Indian Ocean. This notable feat was followed in November by the equally successful Zond 6. It was later revealed that the Zond craft had contained living creatures such as miniature turtles, and that these had not suffered any harm on their 500,000-mile voyage apart from a slight loss in weight. Manned flight seemed more feasible.

Meanwhile in October, the Americans had made a successful manned ten-day Earth-orbital flight of their Apollo 7 spacecraft, a prototype of the vehicle to be used for lunar flights.

On 21 December 1968, the scene was set at Cape Kennedy for the blast-off to man's first flight to the Moon, the Apollo 8 mission. The three-man space capsule stood atop the mighty 363-foot Saturn 5 rocket, crewed by Colonel Frank Borman, Captain James Lovell, and Major William Anders, as the world watched on satellite-relayed television.

At 7.51 a.m. Eastern Standard Time, the five engines of the Saturn 5 first stage successfully ignited, producing a total thrust of 7,500,000 pounds to lift off the 2,750 ton craft. The vehicle accelerated rapidly to a height of 100 miles, jettisonning the first and second stages en route and then entered a 'parking orbit' around the Earth at a speed of 18,000 miles per hour. Two hours 40 minutes later the third stage of the Saturn rocket was fired for a period of $5\frac{1}{2}$ minutes to accelerate the spacecraft to its escape velocity of 24,200 miles per hour and head it towards the Moon.

The firing was so accurate that only one small course adjustment was necessary during the 234,000-mile flight towards lunar interception. During the flight the crew relayed direct television pictures of sights never before seen by man—the Earth as seen from space.

The next crucial point came as Apollo 8 neared the Moon. Because of the effect of the Earth's gravitational pull, the spacecraft slowed down steadily to what was, relatively speaking, 'crawling speed' until it reached a distance of about 205,000 miles where the Moon's gravitational pull exactly

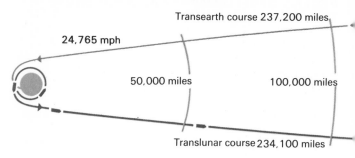

Transearth course 237,200 miles

24,765 mph

Translunar course 234,100 miles

50,000 miles

100,000 miles

The trajectory of the Apollo 8 mission

balanced that of the Earth. Thereafter the craft accelerated under the effect of the Moon's gravity to a speed of 5,750 miles per hour. As Apollo 8 passed round the far side of the Moon, it was necessary to fire the rocket to slow the vehicle down to lunar orbital velocity, about 3,600 miles per hour.

A mistake at this stage could have been fatal, but the crew faultlessly coaxed their craft into an elliptical orbit around

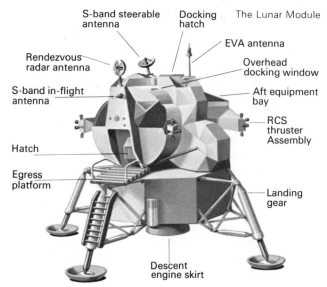

S-band steerable antenna

Docking hatch

The Lunar Module

EVA antenna

Rendezvous radar antenna

Overhead docking window

S-band in-flight antenna

Aft equipment bay

RCS thruster Assembly

Hatch

Egress platform

Landing gear

Descent engine skirt

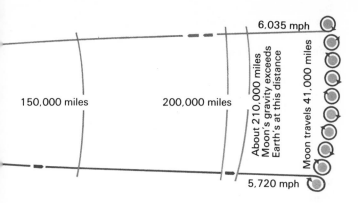

6,035 mph

150,000 miles

200,000 miles

About 210,000 miles
Moon's gravity exceeds
Earth's at this distance

Moon travels 41,000 miles

5,720 mph

the Moon, ranging from a height of 74 to 100 miles, and on the third orbit, the path was altered to a near-circular one at a height of 70 miles. Borman, Anders and Lovell thus spent Christmas Eve making ten orbits of the moon, taking valuable colour photographs and sending back spectacular television pictures. One of the major aims of the expedition was to make a visual survey of possible sites for a future lunar landing, and the astronauts seemed to be particularly impressed by a site in the Mare Tranquillitatis.

Early on Christmas morning, Apollo 8 passed behind the Moon for the tenth time and the rocket motors were once again fired for just over four minutes to accelerate the spacecraft past lunar escape velocity and back towards the Earth.

The only hurdle which remained was re-entry of the Earth's atmosphere at a speed of just under 25,000 miles per hour, much faster than any previous manned spacecraft. The spacecraft had to enter the atmosphere at a very precise angle – too shallow an entry would bounce the craft irretrievably away into space, while too steep an angle would burn it up by friction. True to form, Apollo 8 made a perfect re-entry. The temperature on the outside of the capsule reached 2,800°C as it decelerated. With the aid of parachutes, the craft finally landed gently in the Pacific Ocean only 5,000 yards from the waiting aircraft carrier *Yorktown* after a total journey of nearly 600,000 miles in 147 hours.

Although the Apollo 8 flight was a tremendous technical

and human achievement, it is nevertheless true that it added very little to the sum of our knowledge of the Moon and planets. The flight was important as a necessary preliminary to manned lunar landings.

The two subsequent Apollo missions tested the Lunar Module which was to take men to the surface of the Moon. Apollo 11 in July 1969 was the first attempt at a manned lunar landing and was a brilliant success. The Saturn rocket boosted the Command and Service Module and Lunar Module toward the Moon. In lunar orbit, astronauts Neil Armstrong and Edwin Aldrin crawled through a hatch from the Command Module into the Lunar Module. The Lunar Module separated from the Command Module and descended to the lunar surface, leaving Michael Collins orbiting alone in the CM.

Using its motors to slow its descent and guide it to a clear site, Armstrong skilfully landed the Lunar Module on the Sea of Tranquillity. The astronauts looked out upon a plain pitted with craters and littered with boulders of various sizes. Dressed in his bulky pressurized suit, Neil Armstrong opened the hatch and carefully descended a short ladder to the lunar surface. He found that the surface was firm and that he could walk about fairly easily. The world watched his movements on television. Armstrong gathered some rock samples in case they had to leave suddenly. After lowering equipment from the Lunar Module, Aldrin joined Armstrong on the lunar

(*Left*) The Lunar Module flight path. (*Right*) Schematic drawing of LM linked to Command Module.

The Apollo 11 Moon-landing. Aldrin sets up an aluminium screen to measure 'solar wind', particles emitted by the Sun.

surface. The astronauts gathered samples of rock and dust for distribution to scientists throughout the world for detailed analysis, set up a number of pieces of equipment including a seismograph and an aluminium foil screen to record the 'solar wind', the stream of charged particles emitted by the Sun, and took many excellent colour photographs of the equipment, the Lunar Module and themselves.

Probably the riskiest manoeuvre was the lift off in the Lunar Module from the surface of the Moon and the rendezvous with the Command Module orbiting some 70 miles up. The manoeuvre went perfectly, the two craft docked, and Armstrong and Aldrin rejoined Collins in the Command Module. The Lunar Module was then disconnected and sent off into lunar orbit, and the Command Module returned to Earth and splashdown in the Pacific.

Although the American method of getting men to the Moon may seem complicated, it is in fact the most economical way. To send a rocket directly from the Earth's surface to the surface of the Moon and back would require a rocket at least twice as powerful as the mighty Saturn 5. It is thought that the most powerful Russian rocket, Proton, is considerably less powerful than Saturn (although there are rumours of a much more powerful one under construction), and so the direct lunar flight must be ruled out for some time.

The Russians are likely to tackle the problem in a different way by constructing a manned space station orbiting the Earth which can be used as a launching platform for future lunar and planetary flights. This would have the great advantage that any lunar rocket would only require to be able to accelerate from Earth orbital velocity (in excess of 17,000 miles per hour) to escape velocity (25,000 miles per hour) in order to reach the Moon. Thus there would be plenty of power in reserve for a direct lunar landing.

Besides being a base for interplanetary rockets, the value of a space-station as an astronomical observatory and general scientific laboratory would be inestimable. The Russians have already – in January 1969 – shown the feasibility of this by linking up two large manned craft, Soyuz 4 and 5, in Earth orbit to form a temporary laboratory, interchanging crews before returning to Earth, and it will probably not be more than a year or two before a permanent station is set up with crews regularly changed by rocket.

The Americans followed the Apollo 11 triumph of July 1969 with the Apollo 12 flight of November 1969 and further Apollo flights are planned to take place every six months or so. The Apollo 12 astronauts Conrad and Bean landed within walking

(*Above*) Possible orbiting space station

(*Right*) Part of a radar map of Venus. The dark areas may be mountains.

(*Below*) The link-up of Soyuz 4 and 5. Such techniques are essential in assembling space stations.

Soyuz 5

Soyuz 4

distance of Surveyor 3, a remarkable pinpoint landing. They removed Surveyor 3's television camera and set up a number of experiments before rejoining Gordon in the Command Module. This time the Lunar Module, instead of being sent off into space, was directed at the surface of the Moon. The crash produced unexpected results: the seismograph, instead of registering the shock for a few seconds, showed shock waves to be reverberating for 55 minutes. Scientists have not yet fully explained this curious result.

The spectacular manned space flights naturally attract most public attention, but it must be remembered that at present, and for quite some time to come, the major part of planetary exploration will be carried out by means of unmanned space probes and Earth-based instruments. In fact, many astronomers maintain that too much money and effort is being spent on relatively unproductive space flights at the expense of conventional research.

Some astronomers have been pressing for the construction of a 1000-inch infra-red telescope which, at less cost than, say, a Venus probe, would be much more useful generally. Such telescopes are commonly used at wavelengths twenty times greater than visible light (see page 32) and thus have resolving powers twenty times poorer than their optical counterparts. Hence, even a 1000-inch infra-red telescope could only resolve detail as well as a 50-inch visual instrument. However, such an instrument would be extremely useful for scanning planetary surfaces to note any temperature variations (Venus, for example, has been found to possess localized 'hot spots') as well as studying new regions of the spectrum.

Radar astronomy, whereby signals bounced off planets are analyzed, is another technique which has become important recently. The 59-day rotation period of the planet Mercury was discovered in this way, and American radio astronomers at Goldstone, California, have produced a radar map of Venus showing three zones which are thought to be mountainous. The largest radar telescope at present is the 1000-foot 'dish' formed in a natural hollow at Arecibo in Puerto Rico.

In January 1969 the Russians launched further soft-landing craft towards Venus to help clear up the conflicting measurements made by the last Venus probes, while the

The orbiting astronomical observatory (OAO)
A proposed lunar exploration vehicle

Americans launched two further Mariner craft to Mars in the spring of 1969. America plans an unmanned Mars Orbiter craft for 1971, and a similar craft for Venus in 1972. In 1977, the position of the planets Jupiter, Saturn, Uranus and Neptune will be such that a single space probe could be launched on a path which would take it past all four of them. As the conditions will not be right again for 175 years, an attempt is quite likely.

There is no doubt, amongst astronomers that once manned observatories are established, on space platforms or the Moon's surface, the whole of astronomy will benefit greatly. Several designs for lunar bases have been produced, utilizing solar radiation and nuclear reactors for power supplies, where colonies of scientists could live and work, and experiments are well advanced in producing suitable vehicles to travel over the rugged lunar surface.

Manned flight to the nearer planets may well come about in the next decade, although the more distant planets will have to wait longer. Space travellers returning from other planets will present one potentially serious problem, namely that of contamination. When the Apollo astronauts from the lunar landings returned to Earth, they were kept in strict quarantine for a period of three weeks to exclude the remote possibility that they might have picked up some harmful organic material on the Moon. With journeys to, say, Mars, even stricter precautions will be needed. Quarantine will have to be strictly enforced or the idea of a 'space plague' beloved of science fiction writers, might cease to be fantasy.

It seems likely that other stars have planets, and some of these may support intelligent life. The only means of communication open to us at present is radio, and since the nearest star is over four light-years away, any 'conversation' which might take place would be a long drawn-out affair! Although the hope of success is very slim, attempts have been made to communicate in this fashion. Project Ozma, started in America in 1960, attempted to detect signals from 'other civilizations' at a wavelength of 21 centimetres. Since the hydrogen in space emits radio waves at this wavelength, it would be the logical wavelength on which another civilization might transmit. No success was reported.

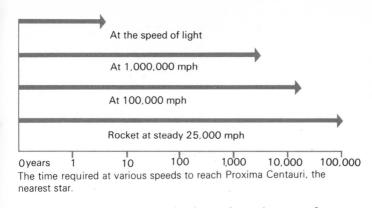

The time required at various speeds to reach Proxima Centauri, the nearest star.

Travel to the stars must lie beyond man's scope for a very long time. Totally new forms of propulsion will have to be invented as present-day chemical rockets are hopelessly inadequate to the task. At a steady 25,000 miles per hour, spacecraft would take no less than a hundred thousand years to reach the nearest star! It is impossible for any object to reach the speed of light, and even if spacecraft could achieve such speed, interstellar journeys would be long-term affairs indeed.

A method of interplanetary communication using digital signals

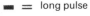

● = short pulse

■ = long pulse

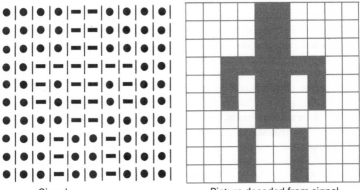

Signal       Picture decoded from signal

# THE AMATEUR AND THE PLANETS

In the past amateurs have contributed valuable observations to the study and exploration of the planets, and even today, when professional astronomers are bringing more and more sophisticated apparatus (not to mention space probes) into action, the amateur, with patience and a reasonable telescope, can still make observations of value.

For some observations, for instance artificial satellites, a pair of binoculars will suffice, but the normal minimum useful size of telescope which can be employed is a 3-inch refractor, and in many ways a 4-inch refractor is far better. The other type of telescope is, of course, the reflector, but in the smaller sizes the refractor is more effective than the reflector. Thus, while a 4-inch refractor is quite useful, reflectors below about 6 inches aperture are not very useful for planetary observations.

It is easy to make a simple refractor which will show many interesting sights. All that is needed is a lens, say 2 inches in diameter and of 30 inches focal length to serve as the object glass, a small, short-focus lens, say 1-inch focal length, for an eyepiece, and two cardboard tubes in which to mount the lenses. Focusing can be done by sliding the eyepiece tube in and out of the object-glass tube. Such an instrument would magnify thirty times, and would give a view of the sky much the same as Galileo saw with his telescope.

Refractors for serious observations must be bought. However, reflectors are much cheaper and can quite readily be constructed by the amateur. Mirrors up to 12 inches in diameter can be fairly easily ground, so that the cost of a powerful instrument can be quite low. For useful observations it is essential that telescopes be well mounted, and that any observation made be accompanied by the following information: name of observer and observing site; date and time of observation (usually in Greenwich mean time, emphatically not British standard time); size of telescope and magnification used; observing conditions and supplementary notes.

There are potentialities for astronomical photography, and particularly in the case of the Sun and Moon, amateurs have produced photographs rivalling professional efforts.

An amateur reflector on an equatorial mount

7 x 50 is a useful binocular size for observing the sky.

A 3-inch refracting telescope on an equatorial mount

145

Domes and rays on the lunar surface

Dark patches in the crater Alphonsus. The arrow shows the position of a volcanic eruption observed by Kosyrev.

←Terminator

## The Moon

Until comparatively recent years, observations of the Moon were carried out almost entirely by amateurs, but with the advent of space exploration, professional interest has been reawakened and powerful instruments brought into play. Detailed photographs taken by such spacecraft as the Orbiter series and by the Apollo astronauts have rendered such amateur activities as mapping surface features redundant, and the amateur wishing to contribute valuable results had to specialize. Virtually the only mapping worth doing any more is the plotting of lunar domes. These low-lying, blister-shaped features are of great interest and, because of their shapes, can only be seen at particular angles of illumination so that telescopes of 6-inch aperture and upwards can be employed here.

Another worthwhile programme of observations is Project Moonhole, which is an attempt to calculate the profiles of the interiors of small lunar craters by combining hundreds of measurements of shadow lengths within these craters at various phases of the Moon. The procedure adopted is to draw the apparent area of the crater filled with shadow and to estimate this as a percentage by measuring the drawing afterwards. This has been found more accurate than estimating the phase by eye at the telescope. The results should then be sent in to the British Astronomical Association, or, in the USA, the Association of Lunar and Planetary Observers. A 4-inch refractor or 6-inch reflector will suffice for these observations, and an 8-inch reflector will cover all the craters on the programme.

Of most interest, perhaps, are the observations of transient lunar phenomena (TLP's). Certain areas on the Moon, notably around the craters Plato, Alphonsus, Aristarchus and Gassendi, tend to show minor coloured glows, brief flashes or obscurations. These phenomena are significant of activity of a minor sort on the Moon and thus of considerable importance. Amateurs can make most useful contributions by watching for these features. A 6-inch telescope can be used, but an 8-inch or 10-inch is of greater value. These phenomena are much more easily seen with a piece of apparatus called a 'Moon-blink', which is basically a holder containing several coloured filters which can be alternated in the field of view.

Some areas of the Moon suspected of change or activity

Theoretical phase of Venus at dichotomy, i.e. 50 per cent illuminated

Actual phase 58 per cent illuminated

0

$1\frac{1}{2}$

$\frac{1}{2}$

2

$\frac{1}{2}$

$2\frac{1}{2}$

0

Intensity measurements

## Mercury

Mercury is hard to find and fairly powerful telescopes are needed to see any of the surface markings. Little can be done by the ordinary amateur, except to note whether or not the Schroter effect (see below) occurs on Mercury as well as Venus.

## Venus

As we saw earlier, Venus is permanently cloaked in dense cloud and little can be seen on its disc. It is rather difficult to observe, being in fact best placed in daylight, when a good equatorial mounting is needed to find the planet at all. Great care must be taken when looking for the planet in daytime to avoid accidentally getting the Sun in the field of view; severe damage to the eyes could result. Near sunrise or sunset, when the planet is visible in a dark sky, it is also rather close to the horizon and the disc is disturbed by atmospheric tremors.

The Schroter effect whereby observed and theoretical phases disagree is well established, but continued observations can be of considerable value in establishing by how much the magnitude of the effect fluctuates from elongation to elongation. The discrepancy can sometimes exceed 6 or 7 per cent of the planet's apparent diameter. The vague dusky patches have yielded no useful information about the planet but should nonetheless be recorded on the intensity scale shown here whenever they are certainly visible.

The cusp caps have not been satisfactorily explained as yet but regular observations may yet lead to a solution. Their position, brightness and extent should be carefully estimated.

The ashen light remains a mystery and careful observations are of considerable value here, as some authorities maintain that the effect is illusory. The most satisfactory way to observe this phenomenon is to place a curved bar (occulting bar) in the field of view of the eyepiece so as to block out the bright light from Venus' crescent and minimize the chance of optical illusions.

Venus is a very unsatisfactory planet to observe in many ways, but it is nonetheless of interest. Instruments upwards of 4 to 6 inches aperture can be employed usefully. The goddess of love hides her secrets well.

Varying aspects of Martian topography sketched by amateurs

## Mars

Mars has perhaps more immediate interest for the observer than any other planet, simply because it is the most Earth-like of the planets and may harbour life. However, it must be admitted that Mars is not easy to observe because of the small size of its disc, the variation in apparent size of the planet, and its long synodic period. This means that really effective observations can only be made near opposition, and close oppositions only occur two or three times in a matter of fifteen to seventeen years. Furthermore, when Mars is at a close opposition it is not well seen from the northern hemisphere of the Earth, where most observers happen to live. Mars does, however, show more detail than Venus and its fascination cannot be denied.

A 6-inch telescope will enable the major features to be studied for a few months around opposition, and a 12-inch can be used for a long and really useful series of results. Important things to watch for are as follows: the appearance of the dark areas, any growth or development of any nature, and their seasonal behaviour. The seasonal behaviour of the polar caps should be noted and particular attention given to the dark collar surrounding the shrinking cap when this is visible. A close watch should be kept for any clouds, blue, white or yellow. These are sometimes seen as obscurations of the Martian surface features.

Colour filter observations can be of notable value, particularly as they may show up any temporary clearings of the 'blue layer' in the Martian atmosphere.

Usually drawings of the planet are made to a standard size, a 2-inch diameter disc prepared beforehand to show any small phase effects and the position and tilt of the poles. The best technique is to study the planet carefully for some time before committing anything to paper. When all the visible detail has been seen, the drawing should then be made as rapidly as accuracy permits, to avoid any confusion in the position of detail due to the planet's rotation. Needless to say, the large details should be recorded first. Descriptive notes are also necessary.

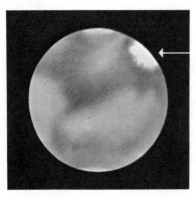

Bright flash indicates presence of cloud

## Jupiter

Jupiter is undeniably the planet which best rewards amateur study—there is plenty of interesting, constantly changing detail visible. As it comes to opposition every thirteen months, it can be regularly studied, and as its apparent size is large, even a 2-inch refractor will show some of the cloud belts. For useful observations though, a 4-inch refractor is the minimum size and a 6-inch is better.

The major task for the amateur is taking transits. When a feature, say a spot on one of the cloud belts, crosses the line

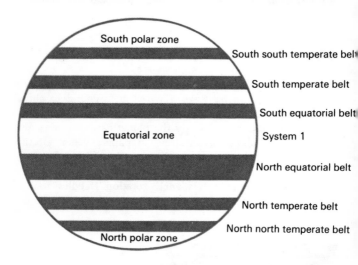

The belts and zones of Jupiter.

joining Jupiter's north and south poles (the central meridian), it is said to transit, and series of observations of transits of features of the cloud belts enable rotation periods to be calculated for these features. Thus any anomalous rotation velocities can be detected and atmospheric currents plotted. Features in or between the equatorial belts are taken to be in system 1 for reference purposes, and features outside this zone are in system 2. A typical (imaginary) sequence of transit observations might read as follows:

1965, December 12. 8-inch reflector, X240. Seeing fair, transparency good.

| G.M.T. | longitude system 1 | system 2 |
|---|---|---|
| 2135 projection on N.E.B. (s) | 237·2 | |
| 2139 preceding end of dark streak in N.T.B. | | 205·5 |
| 2142 white spot in S.E.B. | 241·4 | |
| 2143 central part of streak crossing E.Z. | 242·0 | |

Central meridian

Transit measurements. The spike on the north side of the south equatorial belt is on the central meridian, and thus is in transit. Transit observations have been valuable in attempts to correlate radio waves from Jupiter with surface phenomena.

The time of transit is taken to the nearest minute and the features' longitudes are obtained from tables such as the *Handbook of the British Astronomical Association* after observing is finished. The motion of features past the central meridian is fairly rapid; experienced observers rarely make an error of more than a couple of minutes. Disc drawings should occasionally be done to show the general appearance of the planet; speed is essential because of the rapid rotation. Strip sketches can be made to show the development of features.

## Saturn

Saturn is a splendid sight in telescopes larger than 2 inches aperture and attention is naturally focused on the ring system. However, the serious amateur should devote some time to the globe of the planet. The cloud belts are less pronounced than on Jupiter and usually lacking in activity and thus detail. Consequently transitible features are quite rare, and transit observations, when they can be made, are of considerable importance since our knowledge of the rotation

Features to note on Saturn, with intensity scale. Intensities range from 0 (bright white) to 10 (black sky).

Ring A (3)

North temperate belt (3)

South polar zone (4)

Shadow of globe on rings (10)

Ring B, outer (1), inner (1½)

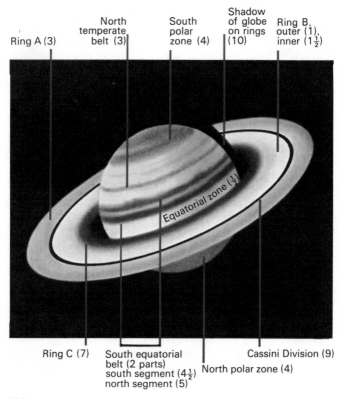

Equatorial zone (½)

Ring C (7)

South equatorial belt (2 parts)
south segment (4½)
north segment (5)

North polar zone (4)

Cassini Division (9)

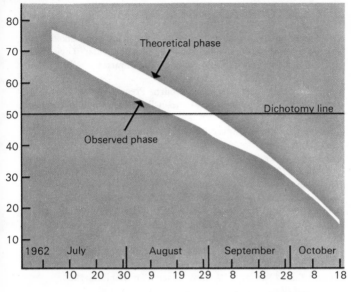

The combined results of members of the British Astronomical Association on the phase of Venus. The discrepancy between observed and theoretical phase was particularly marked in 1962.

period at different latitudes is far from complete. Such features as white spots should be reported immediately.

The shape of the shadows of the rings on the globe, and of the globe on the rings, should be noted, together with intensity measurements of the rings and surface features. Intensity is estimated on a scale of 0 (brightest) to 10 (black shadow) where the outer part of Ring B is assumed to be 1. The position and width of any minor divisions in the rings are best measured with a micrometer, but measuring sketches can yield useful results.

When a star passes behind the rings and/or the planet, the variations in intensity it experiences before being completely hidden can give valuable information on the density of the rings and the planet's atmosphere. Observations of the brightness of the satellites are also useful.

# BOOKS TO READ

The following are a small selection from the many good books available on general astronomy and the planets in particular. They should be available from bookshops and/or public libraries.

*Astronomy* by Patrick Moore. Oldbourne Press, London, 1961.

*Basic Astronomy* by Patrick Moore. Oliver and Boyd, Edinburgh, 1967.

*The Encyclopedia of Space* edited by Maurice Allward. Hamlyn, London, 1969.

*Exploring Other Planets* by P. Moore and H. Brinton. Odhams, London, 1965.

*The Flammarion Book of Astronomy*. Allen and Unwin, London.

*A Guide to Mars* by Patrick Moore. Muller, London, 1965.

*Moonflight Atlas* by Patrick Moore. Mitchell Beazley, 1969.

*Photographic Atlas of the Moon* by Z. Kopal. Academic Press, London, 1965.

*A Planet Called Earth* by G. Gamow. Macmillan, London.

*The Planets* by Patrick Moore. Eyre and Spottiswoode, London, 1962.

*A Star Called Sun* by G. Gamow. Macmillan, London, 1965.

*Stars and Planets* by G. Abetti. Faber, London, 1966.

Apart from the publications of the astronomical societies, two useful general publications available through newsagents are *Sky and Telescope* and *Astronomy Today*.


# ASTRONOMICAL SOCIETIES

The value of joining an astronomical society cannot be too strongly emphasized, particularly for those interested in serious observations. Most national societies have observing sections devoted to the planets where experienced observers correlate observations sent in by amateurs and advise beginners on how to get the best use from their equipment.

The British Astronomical Association, 303 Bath Road, Hounslow West, Middlesex, England.

The British Astronomical Association (Australian Branch), The Observatory, Sydney, Australia.

The Junior Astronomical Society, The Secretary, 17 Chanctonbury Chase, Redhill, Surrey, England.

The Royal Astronomical Society, Burlington House, Piccadilly, London W.1., England.

# INDEX